On the technological foundations of interstellar space travel

Dr. rer. pol. Erik Kolek

On the technological foundations of interstellar space travel

Chronicles of Business Informatics Physics (CBIP)

Volume 3, edition no. 1.0

2024

Imprint

Copyright © 2024 All rights reserved: Dr. rer. pol. Erik Kolek Author of this book and editor of the Chronicles of Business Informatics Physics (CBIP) digital and print version. Publisher: BoD • Books on Demand GmbH, In de Tarpen 42, 22848 Norderstedt, Germany. Print: Libri Plureos GmbH, Friedensallee 273, 22763 Hamburg, Germany.

Scientific citation:

Kolek, Erik (2024). On the technological foundations of interstellar space travel. In: *Chronicles of Business Informatics Physics (CBIP)*. Volume 3, edition no. 1.0. ISBN: 978-3-7597-8523-7.

The Chronicles of Business Informatics Physics (CBIP) consist of cited significant scientific papers and the individually published research articles of the author Dr. rer. pol. Erik Kolek. As an author and an editor of the Chronicles of Business Informatics Physics (CBIP), Dr. rer. pol. Erik Kolek (erster.kontakt@erikkolek.de) is also responsible for editing, translation, typesetting, design (including cover), texts, images and cover picture. The book is produced and published by BoD – Books on Demand, Norderstedt, Germany.

All information and contents in this publication are without legal guarantee. The use of this work and in particular the (experimental) implementation of the information contained therein is expressly at the risk of third parties. The author or editor and the

Kolek, Erik (2024). On the technological foundations of interstellar space travel. In: *Chronicles of Business Informatics Physics (CBIP)*. Volume 3, edition no. 1.0. ISBN: 9783759785237.

Foreword

This book is a groundbreaking theoretical treatise on the technological foundations of interstellar space travel with the research goal of describing and therefore enabling it as down-to-earth as possible. As an introduction, the basics of a theory of everything are described. A quantum medical molecular theory with regard to human body cells is introduced by means of heuristic points of view. Advanced quantum technologies are developed and described in terms of content. The work consists of interesting individual contributions on individual topics. This book is a scientific treatise in the field of business informatics on the one hand and physics on the other. This book is aimed at readers with an interest in the subject.

Following the modern idea of an advanced higher education institution for the exploration of the stars, I have developed innovative program ideas for the alpha development of various quantum technologies based on universal quantum physics principles using a science of everything, which I call business informatics physics – of course, this is a pure *science fiction approach*.

The present forward-looking scientific narrative relates to the technological foundations of a sovereign starship class for light-scaled time-period exploration with a maximum 10c light factor potential, but this currently seems impossible without an appropriately designed star-academic discovery center that is also a museum and whose members are engaged in the content of space-time physics and quantum technologies.

In this book, the technological foundations of interstellar space travel are presented one by one. These begin with a theory of everything, after which the biobed and light resonance tomography (LRT) are presented. This is followed by the presentation of a visor quantum technology in the form of an eye prosthesis designed essentially for humans, which could enable blind people to see again. This is followed by a technological description of the warp drive with a maximum light speed of 10c.

Kolek, Erik (2024). On the technological foundations of interstellar space travel. In: *Chronicles of Business Informatics Physics (CBIP)*. Volume 3, edition no. 1.0. ISBN: 9783759785237.

Following on from this, the nuclear fusion of gases within a reactor based on a neutrino reaction is explained as well as an engineering method for the step-by-step development of innovative technologies. This is followed by a description of the Erik-Kolek-Motor (E-KOMO).

None of the technologies described here are available in the form of a prototype. Only theoretically possible technologies are presented and discussed. The design of prototypes is only a logical consequence that is possible based on the contents of this book. It is important to understand that these are technological foundations or innovative developments that should, at least in theory, make interstellar space travel possible for mankind.

Kolek, Erik (2024). On the technological foundations of interstellar space travel. In: *Chronicles of Business Informatics Physics (CBIP)*. Volume 3, edition no. 1.0. ISBN: 9783759785237.

May 2024.   Dr. rer. pol. Erik Kolek, Diplom-Betriebswirt (FH), M.A., M.Sc.

Kolek, Erik (2024). On the technological foundations of interstellar space travel. In: *Chronicles of Business Informatics Physics (CBIP)*. Volume 3, edition no. 1.0. ISBN: 9783759785237.

Table of contents

Kolek, Erik (2024). On the technological foundations of interstellar space travel. In: *Chronicles of Business Informatics Physics (CBIP)*. Volume 3, edition no. 1.0. ISBN: 9783759785237.

Kolek, Erik (2024). On the technological foundations of interstellar space travel. In: *Chronicles of Business Informatics Physics (CBIP)*. Volume 3, edition no. 1.0. ISBN: 9783759785237.

First section: On the foundations of a theory of everything

Scientific citation:

Kolek, Erik (2024). On a unified theory of everything to explain the entire universe. In: *On the technological foundations of interstellar space travel.* Chronicles of Business Informatics Physics (CBIP). Volume 3, edition no. 1.0.

Erik Kolek (2024)

On a unified theory of everything to explain the entire universe

Summary

The question of whether the universe is a dark or bright universe is easy to answer. It is a dark universe with many bright points or stars. Actually, our universe should radiate brightly, but it can be assumed that dark matter absorbs this bright radiation. But what is the universe and how did it come into being? This is a question that many scientists, such as Albert Einstein and Stephen Hawking, have grappled with. Does the universe have a beginning and will there be an end? Are a big bang and final bang possible, i.e. conceivable? That depends on how you look at it. In this research article, the premise is that our universe formed as a bubble within a giant black hole. This is an expanding bubble that is growing outwards. However, size is relative, just like time. Our universe could also be much smaller than actually assumed. Previous findings remain unaffected by this premise, which means that the assumption of a bubble universe does not contradict existing assumptions. As will be shown, it is very easy to find a possibly valid world formula under this premise. This world formula could help us to gain a completely new understanding of our universe.

On a unified theory of everything to explain the entire universe

So if the universe is inside a giant black hole and everything is structured like a giant bubble, then it is possible that there could be other bubble universes inside this or

Kolek, Erik (2024). On the technological foundations of interstellar space travel. In: *Chronicles of Business Informatics Physics (CBIP)*. Volume 3, edition no. 1.0. ISBN: 9783759785237.

other black holes. The edge of these bubble universes should shine bright red from friction with the dark matter of the outer black hole or its event horizon [6]. Assumptions such as that black holes are actually gateways to other universes thus become more likely, but there is no way into a bubble universe from the outside. Within the bubble universe, stars are distributed in the same way as black holes.

A mathematically most probable world formula and world formula with the simplest explanation for everything could be as follows:

$$\sum c = \sum s = \sum ((h * c^3) / (8 * \pi * G * M * k))$$

This is the Hawking radiation [3, 4, 5], which was written as a sum in order to clarify its relationship to a bubble universe. This rules out the assumption of a flat universe. If light always travels at the same speed, this means that time always passes at the same speed and space can be the same size. This is because the speed is equal to the distance.

In this way, a new model of the origin of universes in physics emerges. At the same moment as a black hole could be created from a stellar explosion, or rather stellar implosion, zero to n new bubble universes are created within the gigantic black hole. The Big Bang could therefore also have been a stellar explosion in the form of a bubble. Each black hole then heats up again until finally another sun is created or ignites and all universes burn up or cease to exist. The stellar furnace is never completely out, but heats up again due to friction until the flash point is reached. The origin of everything is therefore a huge star that became a gigantic black hole through an explosion or rather implosion. In a more precise sense, this is a curvature singularity due to the infinite gravity of the heavy masses of the stars or black holes. The entire universe therefore represents a single curvature singularity according to Albert Einstein [1, 2]. The second form of the singularity without curvature can therefore not exist in nature or physics.

Kolek, Erik (2024). On the technological foundations of interstellar space travel. In: *Chronicles of Business Informatics Physics (CBIP)*. Volume 3, edition no. 1.0. ISBN: 9783759785237.

Within the curvature singularity, any black hole can re-ignite. It is even conceivable that a black hole outside the event horizon [6] has a nuclear fusion like that of a star going on. In other words, there could also be a black hole inside a sun. For example, a black hole could have been raised by a star as it passed very close to it. Spherical ring-shaped stars would therefore be quite conceivable with a black hole at their core.

So what is the unified theory of everything? It is a theory that encompasses various theories, such as Albert Einstein's general theory of relativity [1, 2] but also the quantum theory of matter. This theory includes a comprehensive view of the dimensions of our universe. At least four dimensions are probable, but it is possible that up to eleven or twelve dimensions could exist, especially if two different point events are considered. Experience shows that we can no longer perceive dimensions greater than four with our senses. From the fifth dimension onwards, everything starts to become distorted. This view should look like that of a sphere in which many waves and particles of photons are rattling around wildly. It would be a very curved perception that our eyes would show us. Above all, the quantity of photons would create beautiful plays of light and light phenomena.

The unified theory of everything can also be seen as a kind of matrix. This matrix must contain all previous knowledge about the universe. Nothing must be lost. Otherwise, contradictions would quickly arise, which must be avoided, especially when forming a theory of a unified view of the entire universe. With this improved understanding of the universe, better – i.e. more advanced – space travel should be possible.

According to experience, our rocket spaceships are already obsolete and should be replaced by more modern spaceships. These modern spaceships can be inspired by the television series Star Trek, because the assumption that such spaceships would not be feasible is a rejected hypothesis with the unified theory of everything. So are speeds greater than the speed of light conceivable after all? One possible answer to this is that light has a limited speed, but it should be possible to accelerate very large and very

Kolek, Erik (2024). On the technological foundations of interstellar space travel. In: *Chronicles of Business Informatics Physics (CBIP)*. Volume 3, edition no. 1.0. ISBN: 9783759785237.

small bodies beyond this speed, such as electrons. Just because we have not yet accelerated any body close to the speed of light, the latter statement is not impossible to realize. In the vacuum of our universe, there is virtually no friction that could influence a journey at more than the speed of light. A faster-than-light drive or warp drive would therefore have to be developed. This is a special particle accelerator that is attached to the spaceship with two nacelles. The latter view is known from the series Star Trek Enterprise. The actual spaceship section is therefore elliptical or oval at the front.

The special particle accelerator is therefore constructed in two separate nacelles within which the actual physical reaction takes place. This reaction involves the collision of particles, as is the case inside particle accelerators on earth. The only difference is that these particle accelerators only shoot the particles straight out, creating a small gravitational field effect that moves the closed system in the vacuum of our universe. The system moves faster and faster with each particle bombardment, as there is a lack of friction in our universe. The idea that a particle accelerator can represent a warp drive is both fantastic and ingenious. The closed system moves faster the more particles are fired at it. The latter can only work in the vacuum of our dark universe.

The unified theory of everything can also be referred to as the Einstein-Hawking-Theory. Only the publications of Albert Einstein [1, 2] and Stephen Hawking [3, 4, 5] are relevant in the context of this theory. All other works by other authors can therefore not be used. The bridge between quantum physics and astrophysics is hidden in the works of Albert Einstein [1, 2] and Stephen Hawking [3, 4, 5]. One exception, however, is Werner Heisenberg's uncertainty principle, according to which the position or speed of particles can be determined. It should be noted that just because very large as well as very small bodies can move does not mean that time has to pass. Time is therefore an illusion; we only measure it with our clock hand machines. So if there is no time, then there can be no free will. This realization is far-reaching because then everything in the universe would already exist and we wouldn't have to worry about anything. The latter is consistent with the statements made by Stephen Hawking in his books, as he also doubted free will. We are apparently what the universe pretends we are (if we want to be).

The assumption of bubble universes within black holes is a more specific general relativity of everything. Accordingly, each bubble universe has its own model-based reality that our mind provides for us. Albert Einstein's well-known formula [1, 2] $E = mc^2$ applies to stars and this formula should also apply to black holes in the opposite sense, i.e. $-E = -mc^2$. In any case, the curvature of light was proven by a photograph of a solar eclipse. In the same way, Edwin Hubble's finding that our galaxies in our universe are moving away from us and appear red-shifted remains valid. Likewise, God does not play dice as Albert Einstein [1, 2] said, because there is no (statistical) coincidence in our universe, but at most the (esoteric) coincidence. In our universe, which could exist as a bubble in a black hole, it is possible for every present to begin anew again and again. The latter is a mental reference to a closed string. There is causality but no time in the black bubble universe.

We remember the system of the particle accelerator 2.0 (warp drive), whose deceleration can be described as $E = mc^2$ and whose acceleration as $-E = -mc^2$. Supergravity arises within the warp bubble system. The optimal spaceship has a spherical shape and a ring on the equator, namely a ring-shaped particle accelerator. The latter is modeled on an electron that could be accelerated to faster-than-light speeds. The Magnus effect or Bernulli effect must be taken into account here, in which currents outside the spaceship cause it to rotate in a spherical spaceship. The use of electromagnetic fields inside the particle accelerator 2.0 improves this even further. Interstellar (not intergalactic) space travel thus appears increasingly likely. The Lorentz factor is negligible here, as this was only assumed for light by Albert Einstein [1, 2].

If enough black matter collapses to form a black hole and it heats up due to its extreme mass and extreme angular momentum, a huge explosion or implosion can occur,

Kolek, Erik (2024). On the technological foundations of interstellar space travel. In: *Chronicles of Business Informatics Physics (CBIP)*. Volume 3, edition no. 1.0. ISBN: 9783759785237.

known as the Big Bang. A black hole is an extremely heavy mass that rotates around its own axis at approximately the speed of light. It is a perfect sphere, as a 6th dimension exists in space-time. This spherical shape is transferred to the universe as a time bubble when the black hole explodes. The expansion should correspond to the angular momentum of the exploded black hole. This should create a time bubble that moves extremely fast, rotates with the same angular momentum of the black hole and expands. The expansion generates heat at the edge of the universe and this heating should be visible at the edge of the time bubble as a glow or bright shimmer. This corresponds to Hawking's radiation [3, 4, 5] and his later assumptions. Several universes can be created simultaneously, although the word time should be used with caution in this context and rather in the sense of a constantly occurring natural phenomenon in a superuniverse that corresponds to a perfect sphere.

Every entire black universe arises from the primordial universe consisting of supermassive anti-matter. A chaos theory of the supermassive primordial universe states: No universe arises until: (supergravity) sg entire universe = sg primordial universe. Moving anti-matter particles react through friction and a shock or explosion wave is created, resulting in universes that are separated by their event horizons (edges) [6]. The following applies: sg entire universe > sg primordial universe. Universes can theoretically disappear again as soon as: sg entire universe < sg primordial universe, but impossible because sg grows infinitely?: sg primordial universe < sg_1 < sg_2 < sg_3 < sg_n, each sg stage is created by a mass-dependent event horizon [6], at some point mass for sg primordial universe could be used up and sg primordial universe disappear, then sg_1 etc., first signs would be that black holes in the interior stagnate in growth or even disappear.

Is it possible that everything ends with a final bang? Yes, this possibility exists, because it is conceivable that bubble universes could burst directly and thus evaporate just as they were created. The latter could happen due to gravitational effects. It is also

Kolek, Erik (2024). On the technological foundations of interstellar space travel. In: *Chronicles of Business Informatics Physics (CBIP)*. Volume 3, edition no. 1.0. ISBN: 9783759785237.

conceivable that another bubble universe could form and collide with ours or overlap our black universe.

Despite everything, the present unified theory of everything is still incomplete. Individual contributions have been researched, which is due to the theories and interactions themselves. There are still gaps. The theory will never be fully explainable. There is no correct world formula, but only one form of the world formula as described in our example above. The speed of light could represent the sound barrier of the universe. This wall is, so to speak, a wall of light or a wall of time that could be broken through. In a sense, this would also make time travel possible, as all events at any location could be visited by people in spaceships. The whole thing should be taken with a grain of truth.

To return to the more specific general theory of relativity of everything, it must be said that Albert Einstein [1, 2] had already reached a first milestone in the development of the theory. He accurately described the gravitational field with a series of equations. Observers can imagine the sum of the gravitational fields in our universe as a supergravitational field. This supergravitational field rubs against the dark matter of the external black hole as a force. Every black bubble universe therefore has a supergravity field. Friction generates heat at the outer edge of our bubble universe. This high temperature T is reflected in Albert Einstein's well-known equation [1, 2] extended by T because $E = Tmc^2$. Bubble universes and black holes therefore have event horizons [6] in common.

In his books, Stephen Hawking also referred to these bubble universes that I introduced as baby universes. The difference lies in the assumed size of these universes, because as the name suggests, baby universes inside black holes are very small, while bubble universes are very large. Which form of universes are therefore more likely to be found – baby universes or bubble universes? This question will be answered below.

Kolek, Erik (2024). On the technological foundations of interstellar space travel. In: *Chronicles of Business Informatics Physics (CBIP)*. Volume 3, edition no. 1.0. ISBN: 9783759785237.

Both forms of universes, bubble universes and baby universes, could be found inside black holes, where they are created by a blub or big bang. How exactly this blub comes about will be explained later. This is about the discussion of whether baby universes or bubble universes exist in physical reality. The blub creates the universe, which is always a bubble universe because its size is relative. The name baby universe can refer to its age. Young bubble universes could also be called baby universes. Baby universes can therefore also represent bubble universes, because these could be very young and very small. Our bubble universe, on the other hand, is very large and very old about 15 billion years as we assume. It is important to mention that our black bubble universe is still expanding, which means that the bubble is still enlarging.

So how does a blub form inside a black hole? This is an effect in the gravitational field of the black hole, because it is so extremely powerful that it could lead to a division of the gravitational field. A cavity could form within the black hole, which we call a bubble universe. The latter has its own gravitational field, the supergravity field. The bubble is an effect that arises because unimaginable temperatures could occur inside the black hole. It is like a saucepan on the stove whose water forms bubbles when heated. This simple physical effect can also be used to explain the bubbles inside the black hole. It is therefore conceivable that a bubble universe like ours was created in one fell swoop and has been expanding ever since. The latter is consistent with today's assumptions in physics according to experience.

The unified theory of everything also describes a completely new approach to how our universe could have come into being. It is an extension according to which the Big Bang could have taken place as a bubble inside a giant black hole. This assumption allows new far-reaching insights into space-time. This is because both space and time within a black hole behave differently than if it were simply a universe. The bubble universe therefore has a more specialized spacetime that should be subject to different physical laws. This mainly concerns the question of whether it should actually be possible to move faster than light. Until now, it was assumed that only

Kolek, Erik (2024). On the technological foundations of interstellar space travel. In: *Chronicles of Business Informatics Physics (CBIP)*. Volume 3, edition no. 1.0. ISBN: 9783759785237.

light could reach a speed of 300,000 km/s and that a spaceship could only reach this speed approximately. But it is conceivable to travel faster than 300,000 km/s inside a bubble in a black hole, because the laws of physics are different here. At these speeds there should be extreme friction within the vacuum of our universe and there should be a small but existing super-gravitational field around the spaceship. The latter describes the generally known or popular concept of a warp bubble. A warp bubble is a bubble within our bubble universe in the black hole that wraps around the spaceship. In purely physical terms, it should therefore be conceivable to realize a warp drive.

Vibrations of black matter could occur within a black hole, which could lead to a blub similar to those of earthquakes. These vibrations would have to be very strong, so that a division of the gravitational field could occur in their center. This would create the blub, at least in theory. The blub could be visualized as a bubble forming in a hot tar formation. This bubble would be unimaginably hot at its edges and well tempered on the inside. A big bang in the form of a bubble formation should therefore also be conceivable without extremely high temperatures inside. A cooling of our black bubble universe should therefore seem superfluous. This cooling would have occurred directly and not over the course of billions of years. So a blub or Big Bang need not necessarily have been so hot. A blub as an assumed big bang could therefore be quite possible. The consistency of the bubble would then depend on its internal composition. A uniformly heated bubble on the inside should be very stable on the outside. A sudden bursting of our black bubble universe should therefore be ruled out for the time being. The latter could be figuratively identical to the existence of a hot tar bubble.

The unified theory of everything unifies the strong force, the weak force, electromagnetism and gravitation. The latter is possible because it is a special form of a general theory of relativity of everything. This includes Albert Einstein's general theory of relativity [1, 2] as well as the quantum theory of matter. Both very large and very small bodies are described in their respective physics. Our black bubble universe

Kolek, Erik (2024). On the technological foundations of interstellar space travel. In: *Chronicles of Business Informatics Physics (CBIP)*. Volume 3, edition no. 1.0. ISBN: 9783759785237.

as a very large body contains many very small bodies according to its nature. Inside are placed uncountable stars and black holes, which have their own physical states. In this context, entropy refers to the prevailing temperature T as well as to the existing chaos. Nothing happens by chance but almost as planned within the chaos theory. An emergence of life within the black bubble universe seems hardly conceivable without some chaos. Very small bodies such as living beings could be an indication of a functioning view of the world. In any case, we humans will still play a far-reaching role in this black bubble universe of ours. The latter should only succeed if we stop our current conflicts and warlike actions and learn to act together again. Only together is it possible to explore the universe and all the wonders of nature it contains. A unity of mankind is entirely in the spirit of a unified theory of everything.

References

[1]   A. Einstein (1905). Ist die Trägheit eines Körpers von seinem Energiegehalt abhängig? *Annalen der Physik* 18(13), pp. 639–641.

[2]   A. Einstein (1916). Die Grundlage der allgemeinen Relativitätstheorie. *Annalen der Physik* 354(7), pp. 769–822.

[3]   S. W. Hawking (1974). Black hole explosions?. *Nature*. 248 (5443): pp. 30–31.

[4]   S. W. Hawking (1975). Particle creation by black holes. *Communications in Mathematical Physics*. 43 (3): pp. 199–220.

[5]   S. W. Hawking, M. J. Perry, and A. Strominger (2016). Soft Hair on Black Holes. *Phys. Rev. Lett.* 116, No. 23, 231301.

[6]   K. Schwarzschild (1916). Über das Gravitationsfeld eines Massenpunktes nach der Einsteinschen Theorie. *Sitzungsberichte der Königlich Preussischen Akademie der Wissenschaften* 7, pp. 189–196

Kolek, Erik (2024). On the technological foundations of interstellar space travel. In: *Chronicles of Business Informatics Physics (CBIP)*. Volume 3, edition no. 1.0. ISBN: 9783759785237.

Table of contents

In this research article, a unified theory of everything is developed. This is a possible unified theory of everything. There could also be other theories. However, in this theory, our universe was created by a blub or big bang due to the gravitational field changes within a giant black hole.

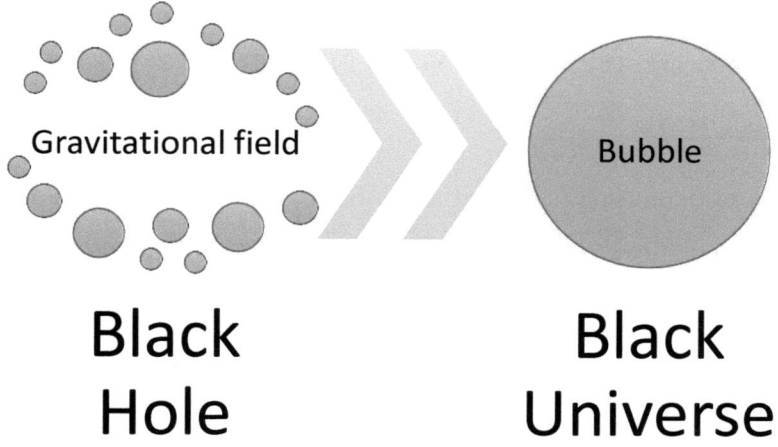

Figure 1. Table of contents illustrated by the importance of a unified theory of everything.

Second section: On the basics of light resonance tomography, a sighting quantum technology and energy pulse technology

Scientific citation:

Kolek, Erik (2024). An advanced molecular theory of body cells as the basis of quantum medicine and development of a light resonance tomography imaging device. In: *On the technological foundations of interstellar space travel*. Chronicles of Business Informatics Physics (CBIP). Volume 3, edition no. 1.0.

Erik Kolek (2024)

An advanced molecular theory of body cells as the basis of quantum medicine and development of a light resonance tomography imaging device

Summary

Beginning with an advanced molecular theory of body cells that sheds light on the RDNA complex structure of living organisms, statements about microbes and environments within planetary atmospheres are also opened up. This foundation of quantum medicine serves the development of an advanced purely medical photon radiation technology. This is light resonance tomography based on Albert Einstein's electrodynamized photoelectric effect, which as a consequence allows a computer and information system-supported biobed and thus a hospital ward for general quantum physicians and their patients. This should be the beginning of the next century in medicine for all kinds of living beings. Of course, Erik Kolek is not a physician and therefore all information is to be questioned.

Kolek, Erik (2024). On the technological foundations of interstellar space travel. In: *Chronicles of Business Informatics Physics (CBIP)*. Volume 3, edition no. 1.0. ISBN: 9783759785237.

An advanced molecular theory of body cells as the basis of quantum medicine and development of a light resonance tomography imaging device

Based on an advanced molecular theory with regard to living cell bodies, quantum medicine can be derived in general terms using Euclidean geometry. The following purely mathematical assumption therefore always applies to a single body cell:

(1) $RNA_{ik} + DNA_{ik} = 1$.

A single somatic cell of a living organism therefore always consists of a certain relationship consisting of a multidimensional complex of ribonucleic acid (RNA) and a multidimensional complex of deoxyribonucleic acid (DNA). This relationship can vary from cell to cell. Pathological changes in the cell can occur if the RNA_{ik} as a whole takes up a larger proportion than the DNA_{ik} as a whole in the cell complex. According to the theory of heredity, this can result in damage to the overall structure of the body cell and be passed on. This also expresses the diversity of every living body cell. The factors m and n in (1) must be integrated in order to take into account the multiplicity of a single multidimensional somatic cell. From this follows purely mathematically:

(2) $mRNA_{ik} + nDNA_{ik} = mn$.

In accordance with Euclidean geometry, according to (2) a somatic cell can be imagined mathematically simplified as a two-dimensional circle or physically as a multidimensional ellipsoid (3). If the size m of the RNA_{ik} is increased externally, the product mn and thus the somatic cell as a whole also grows. This also generally explains the growth process of a single body cell. Therefore, as a piece of quantum medical advice, an (artificial) non-uniform change in the factors m and n should always be avoided in order to be able to counteract pathological changes in the body cell, for example, to possibly cure cancer with a pill that strengthens or balances the DNA complex. In the following, the aforementioned two-dimensional surface of a circle or multidimensional surface of an ellipsoid according to Gauss will be considered as representative of a body cell. Instead of the product mn, we now introduce the invariant ds^2 as Albert Einstein (2009) methodically does.

(3) $ds^2 = RNA_{11}dDNA_{11}^2 + 2RNA_{12}dDNA_{11}dDNA_{12} \ldots + RNA_{44}dDNA_{44}^2$ (Einstein, 2009, p. 59)

Using (3), the newly derived complex RNA-DNA structure of a somatic cell is easy to understand. This is a four-dimensional space-time continuum of a somatic cell that can also assume non-Euclidean shapes. Non-Euclidean shapes are taken into account by considering the surface using Gauss and are therefore possible. For the general case of an (approximate) Euclidean complex RNA-DNA structure of a somatic cell, the following expression applies.

(4) $ds^2 = dRDNA_1^2 + dRDNA_2^2 + dRDNA_3^2 + dRDNA_4^2$ (Einstein, 2009, p. 59)

Based on the newly derived complex RNA-DNA structures (3) and (4), it quickly becomes apparent that (artificial) changes in the body cell structures must always be pathological, whereby RNA and DNA are always firmly linked and therefore formulated as RDNA, whereby a positive RNA complex change always causes a negative DNA complex change and, conversely, a positive DNA complex change always causes a negative RNA complex change within the body cell. This means that there is always a death of the affected body cells (cancer attack) and, conversely, always a revival of the affected body cells (cancer cure). Long-term negative consequences on the RDNA complex structure, particularly through heredity, cannot currently be treated with quantum medicine. However, a corresponding pill to strengthen the DNA complex with a balancing effect on the RNA complex (cancer cure) seems conceivable, but this pill is unlikely to be able to eliminate any inherited DNA complex damage, as the complex RDNA starting structure in the body cell is already damaged. Quantum medicine as an advanced type of medicine no longer distinguishes between human medicine and veterinary medicine due to the complex RDNA structures of body cells generally present in every living being according to

Kolek, Erik (2024). On the technological foundations of interstellar space travel. In: *Chronicles of Business Informatics Physics (CBIP)*. Volume 3, edition no. 1.0. ISBN: 9783759785237.

nature. Quantum medicine is therefore a standardized medicine for every type of living being.

Quantum medicine also deals with the smallest living organisms that can come into contact with individual body cells. In the physical nature of the space-time continuum, these smallest living beings are generally all types of microbes that can occur on all types of planets and are subsequently referred to collectively as microbes for quantum medical reasons. At this point, we should refer directly to the medical microbiologist Louis Pasteur, who famously said that only the environment was sick and that the microbe actually had nothing to do with it. This statement is now to be tested by means of theoretical quantum medicine in order to uncover any existing dangers for humans in the colonization of planets.

Microbes within planetary atmospheres. Most likely, Brownian motion (Einstein, 1905, 1906) has a decisive influence on the movement of microbes within heated gases and liquids. To simplify matters, it should be assumed that the microbes have a spherical shape and can move and rotate within a planetary atmosphere. With regard to the size of the microbes, it should actually be clear that their size cannot be the same as the size of the air particles (e.g. oxygen O_2), because otherwise the microbes themselves would not breathe and would therefore not be viable if the definition of breathing life forms is generally followed. Microbes are therefore larger than the gas particles, which is why they can move between the gas particles and even use their movement impulses for acceleration, whereby the necessary oxygen O_2 is probably absorbed for the most part via their skin surface. However, microbes cannot survive for long within planetary gases, as they need food in the form of cell nuclei (RDNA) just like all other breathing life forms, so it would of course be conceivable that they could eat each other when there is a lack of food, but they could also equally infest all types of living organisms such as plants, animals and humans, because microbes feel most comfortable in warm liquids such as water and an aqueous reddish metal hydroxide (metal water solution), which, thanks to medical quantum chemistry in

Kolek, Erik (2024). On the technological foundations of interstellar space travel. In: *Chronicles of Business Informatics Physics (CBIP)*. Volume 3, edition no. 1.0. ISBN: 9783759785237.

conjunction with the RDNA complex cell structure, makes a new promising healing potential conceivable with regard to physical function research in vitro with cell samples placed in a certain metal hydroxide with the addition of certain nutrients such as amino acids. So as long as microbes are moving in bodies outside of warm liquids, they pose no danger to the environment; Pasteur is probably still correct here. Otherwise, all types of living organisms would die directly in the planetary atmosphere, precisely because they breathe. Although some microbes are continuously inhaled, they can hardly settle in the environment if they are filtered out of the gas atmosphere on their own, as the environment has sufficient defense functions.

Milieus within planetary atmospheres. The situation is different when microbes suspended in warm liquids, as described in Brownian motion by Einstein (1905, 1906), are inhaled at an accelerated rate through milieus from other milieus, because in these exhaled cell samples the microbes have already multiplied significantly through mitosis. Within confined space-time continua (as given in a smaller space), infection with microbes can therefore occur, even within various metric time periods, since here Brownian motion lasts longer due to heat within the limited four dimensions $dx_{i=4}$. Incidentally, the same principle can be applied to any solar system, but I don't want to go into this in detail as it would also take me off topic. So after the microbes have entered the environment via the mucous membranes, which generally absorb and separate the particles, and are still really inside their warm liquid droplet, because this also protects them from direct access by the defense system (unlike microbes that move back and forth between gas particles), they begin to multiply as soon as the hydroquant is transported into a body cell for temperature regulation. Once inside the body cell, the water particle is slowly absorbed until only the microbes remain. These microbes are then in the vicinity of the cell nucleus and try to penetrate to this nucleus, biting their way forward bit by bit until they have arrived there. Now the body supplies the infected microbial cell with nutrients, which could lead to pathological multiplication of the microbes on the one hand and the growth of the cell

Kolek, Erik (2024). On the technological foundations of interstellar space travel. In: *Chronicles of Business Informatics Physics (CBIP)*. Volume 3, edition no. 1.0. ISBN: 9783759785237.

on the other. However, the latter depends on the type of microbe. The growth of the cell also requires more nutrients, but these are intercepted by the microbes, which is why the cell gradually dies and turns black. Black dead somatic quantum cells are microtumors that are normally eliminated by the phagocytes of the immune system. If these heavier black hole cells, which contain a lot of iron, zinc and magnesium, for example, are not broken down early enough by the immune system, pathological growth occurs until the tumors are visible or noticeable and have to be surgically removed. The degradation of the black hole cells and the metal elements they contain causes the otherwise clear water to turn red at quantum level. In quantum medicine, blood can therefore be regarded as an aqueous reddish metal hydroxide (metal water solution).

Microbes do not like to stay in the vicinity of black hole cells, which is why they migrate to the next body cell until at some point they suddenly encounter a quantum body cell that has already adapted to the microbe infestation, because the scavenger cells have decoded the microbes through their work and produced corresponding anti-body cells in the closer four-dimensional space-time environment. The necessary information for this is passed on via the metal hydroxide using the RDNA complex code of the microbes. Known microbes are quickly decoded correctly, which is why not every microbial infestation leads directly to a disease. Unknown microbes are more difficult to decode, which can lead to errors in the RDNA complex code learned by the immune system, allowing the microbes to spread for the time being. Immunized body cells are protected against microbial attack because they have acquired a better DNA complex structure and already know the microbe with its RNA complex structure, so to speak. If the microbe tries to bite into such an anti-body cell, it is dissolved by the modified acid (DNA) because its surface can no longer offer any resistance. Ultimately, an RDNA complex structure of an environment is a self-regulating body system against microbes that appears so extremely diverse and modular that no external intervention seems necessary at present, not even in quantum medicine, to support this automatically learning immune system. Although there are

Kolek, Erik (2024). On the technological foundations of interstellar space travel. In: *Chronicles of Business Informatics Physics (CBIP)*. Volume 3, edition no. 1.0. ISBN: 9783759785237.

apparently physically medically justified cases in which the introduction of various chemical elements appeared to be effective, this is also merely a control and learning mechanism of the body's immune system of the environment, as it is not known whether this effectiveness would also have taken place without the introduction of various chemical elements. Countless experiences of the body's defense system always remain unknown even to quantum medicine, because the number of types of microbes, even within microbial strains, remains non-Euclidean infinite. It is now also clear that the introduction of chemical substances into the body cell for purely psychological protection against a single specific microbe seems pointless, and may even be hostile to the body cell, since, on the one hand, infection with precisely this single microbe is actually already physically impossible in quantum medicine due to the diversity and modularity of the nature of microbial cells. It could even promote black hole cells and thus self-generated pathologies. This also directly answers the question of when quantum medical injections should be made into body cells. The answer is: from the first appearance of symptoms and not before, because at this moment the body cell is healthy and should remain so. To summarize, it can be said that Pasteur expresses the truth with regard to the milieu, but at present no organic chemical protective mechanism must or can be introduced into the body cell that establishes a physically really existing precaution, probably not even in quantum medicine, at least not for all types of microbial cells that occur in nature. To assume and even strive for this is therefore an illusion.

In modern quantum medicine, injections are generally carried out differently than in classical medicine, especially for subcutaneous injections, metallic needles appear to be completely superfluous, as they cause pain when injecting the patient with medication, for example. The latter sometimes even causes psychological pain for some quantum physicians. Modern injection devices are therefore designed completely differently to what people are used to today, i.e. no more plastic cylinders with a needle tip at the front. Above all, the injection devices are no longer operated manually by medical staff such as physicians, but by pressing a button, which can even be done by the patient themselves if they wish.

Inside the injection device, which I call a quantum medical hydrospray, there is a small triangular container for holding the medical hydrate. The tip of the triangular container points downwards towards the exit plane, the diameter of which can be adjusted in micrometers from the outside in a circle, depending on the desired depth of the four-dimensional space-time area into which the medical hydrate is to be injected. The acceleration of the medical hydrate replaces the metallic needle and is generated by a compression process in the hydrospray. Pressure and density naturally play a role in compression. It is generally accepted that

Compression density μ = density Ω – pressure Δ.

For the density Ω, we assume 1 kg per 1 liter of medical hydrate, i.e. 1000 kg/m^3 or 1 g/cm^3 as the SI unit. From a physical and mathematical point of view, it must be taken into account that no gas particles such as oxygen O_2 may be introduced into the hydrospray for quantum medical and compression-related reasons. For this reason, the hydrospray must be applied directly to the puncture site, otherwise the air pressure would have to be taken into account. The puncture site is a point coordinate that is simply targeted using a laser pointer. A 1 cm hydrate column is to be introduced into the body cells via an arbitrarily selectable skin site, i.e. 0.001 bar or 100 Pascal (Pa) pressure Δ is required as a common SI unit. Numbers are only used here for visualization, as arithmetic has already been disproved by proving Fermat's theorem $2\mu^{\sigma} \neq \mu^{\sigma}$ for any number. This means that numbers are merely mental concepts that do not occur in the physical reality of nature. Nevertheless, numbers are suitable for explaining relationships in mathematical theorems with regard to physics. We obtain:

Compression density μ = 1 – 0,001 = 0,999.

The hydrospray must therefore have a high compression density so that the hydrate column can be introduced with a targeted shot at the push of a button. A hydrospray

Kolek, Erik (2024). On the technological foundations of interstellar space travel. In: *Chronicles of Business Informatics Physics (CBIP)*. Volume 3, edition no. 1.0. ISBN: 9783759785237.

is therefore similar to a medical atomic syringe, in which the liquid matter undergoes a very high acceleration as soon as the atmosphere is equalized through the opening of the outlet cap. The question naturally arises as to how this compression can be generated. The answer is pneumatics. Various types of pneumatic compression devices already exist in medicine today, but they are large and mostly compress air, so it is to be expected that the first actual hydrosprays will have to be connected to a trolley via a hose. Later, thanks to nanotechnological miniaturization of the first prototypes, the idea of a handy hydrospray can really be realized. Ballistic gelatine should be used to carry out initial tests of the hydrospray; please do not carry out any animal experiments, as this is no longer common in quantum medicine research and practice. Now a hydrospray can also be imagined as a medical water pistol, not only by the specialist medical staff, which has a corresponding tank and can be used several times on different patients without taking any medically relevant risks, for example hygienic risks for the patients. Quantum medical treatment saves time and money and is pain-free and safer, which should of course please any patient who suffers from needle phobia. This much more pleasant way of injecting will certainly take away or at least alleviate the fear of injections for many patients, and not only within reference body classes (spaceship classes).

As a braking effect is produced by the body cells immediately upon injection of the medical hydrate, the injection depth and hydro mass must be determined in advance. Corresponding experience can be gained using the ballistic gel, but the distribution of the hydroquanta in the body tissue is expected to be different from puncture with an ordinary plastic syringe. While with such a syringe the medication is distributed around the needle in the form of bubbles, this distribution must be considered differently with the hydrospray. The acceleration of the hydrate decreases unevenly after passing the first skin cells, because one person has slightly thicker skin than another, and the same is true for other living beings. In every general case according to physics, a mushroom-like distribution function of the hydrate should result, whereby the style of this function can be of different lengths.

As soon as the aforementioned nanotechnological innovations enable mobile pen-like hydrosprays, a lithium battery or, better still, a multidimensional galvanic element connection will be located inside the hydrospray. Maxwell's field equations of electrodynamics can be used in the development of multidimensional galvanic element compounds, and electromagnetism and electrostatics in particular should probably play a role here. This should make a truly long-lasting battery possible and keep the hydrospray ready for use for as long as possible. Mobile hydrosprays have the disadvantage over stationary trolley variants that they will have a smaller tank, as the battery will also take up space in three dimensions in Newtonian terms. Just as the battery can be improved, it is also conceivable in mechanical engineering that effective pneumatic compression could be replaced by more efficient electrodynamic compression. The *anti-general theory of relativity* derived by Erik Kolek with regard to dynamic gravitational magnetic fields provides a suitable physical foundation for the development of quantum technologies.

Electrodynamic compression can also be assumed to occur within the body's own cells of living organisms, as fewer but still individual photons come into contact with these body cells. In quantum medicine, skin cancer is therefore always an oncological indication of pathological tumor growth processes inside the respective cell body. The latter is also related to the fact that the assumption that light rays only hit the skin is now an outdated Newtonian way of thinking. Photons already penetrate the outside of the skin at the speed of light and can even travel on their curved lines through the body to the other inside of the skin, which the photons penetrate repeatedly. So there could be a kind of (partially) phosphorescent but invisible radiation of the body of living beings from the outside to the inside and from there to the outside again, therefore all kinds of living beings should not expose themselves to direct sunlight for too long without appropriate radiation protective clothing due to the risk of cancer inside and not only of the skin in the case of prolonged light absorption of phosphorescent bodies.

Kolek, Erik (2024). On the technological foundations of interstellar space travel. In: *Chronicles of Business Informatics Physics (CBIP)*. Volume 3, edition no. 1.0. ISBN: 9783759785237.

With the help of this hypothesis about body afterglow, it is also conceivable that photosynthesis could also take place inside living beings such as animals and humans, albeit with less available light than in plants, but nevertheless continuously during light absorption. Therefore, it could also be that during darkness, for example at night, more weight is gained when snacking, as the body fat cells could require additional photons, i.e. radiation energy, for their dissolution. This photosynthesis of living organisms in general could also make their digestive processes easier to explain, which in turn could lead to a better understanding, diagnosis and treatment of diseases. How else would nutrients be broken down into their atoms if Albert Einstein's (1905b) electrodynamized photoelectric effect could not also play a role here? It could therefore be assumed that a certain supply of light could generally promote digestion, and that the microorganisms living there, such as the microbes in the intestine, could also benefit from this light emission and thus perform more efficiently. For this reason, an infrared lamp could be recommended instead of a hot water bottle as an old, but actually new, quantum medical home remedy for intestinal pain and other intestinal complaints in order to increase the acceleration of residual element removal and to better dissolve any existing constipation. Since darkness and lightness usually alternate in a reference body's own space-time continuum, quantum medicine should be able to provide a wealth of new insights thanks to the photoelectric effect (Einstein, 1905b) inside living beings.

Figure 1 shows that the innovative idea of phosphorescent bodies and photosynthesis inside living organisms based on Albert Einstein's (1905b) photoelectric effect is not unfounded; it shows an X-ray image of a hand. This X-ray image, which is no longer black and white but can now be digitized in color, was taken using a simple quantum technological light resonance tomography (LRT), the technical explanation and implementation of which follows in the next paragraph. Physically, LRT only works in darkness due to the phosphorescent bodies; living beings appear to radiate (partially) reddish light from the inside outwards due to the gravitational field, which should make it possible to determine the resonance of the light rays because some of

Kolek, Erik (2024). On the technological foundations of interstellar space travel. In: *Chronicles of Business Informatics Physics (CBIP)*. Volume 3, edition no. 1.0. ISBN: 9783759785237.

the photon radiation is also absorbed. A low modern photon radiation (instead of a high classical electron radiation) was set on the X-ray image, but the finger bones are still partially recognizable. Pathological body cells with a higher density, such as tumors, could appear darker and of a different color on such LRT images. Corresponding Fourier series, named after Joseph Fourier, can also make these medical Einstein images easier to differentiate using LRT, especially during the sectional camera images for better color scaling, which should make X-ray devices, computer tomography (CT) devices and magnetic resonance imaging (MRI) devices technologically obsolete. Quantum medicine also no longer requires toxic contrast agents, as these usually contain elements that are expressly forbidden to be injected into living bodies, such as radiating elements of any kind, metals and heavy metals such as gadolinium (also not as chemical compounds etc.), as these do not correspond to the body cell elements of living beings. Only elements that are similar to body cells may be introduced into living beings from the outside without quantum medical concerns, i.e. injected or ingested with food, i.e. no chemical aluminum compounds and so on, but for the time being only chemical carbon-silicon compounds (probably only chemical silicon compounds in terms of the reference system). Organic quantum chemistry as a component of quantum medicine must be further derived and approximated in order to determine safe quantum medical element compounds, all of which can be used within living beings without any health risks.

Kolek, Erik (2024). On the technological foundations of interstellar space travel. In: *Chronicles of Business Informatics Physics (CBIP)*. Volume 3, edition no. 1.0. ISBN: 9783759785237.

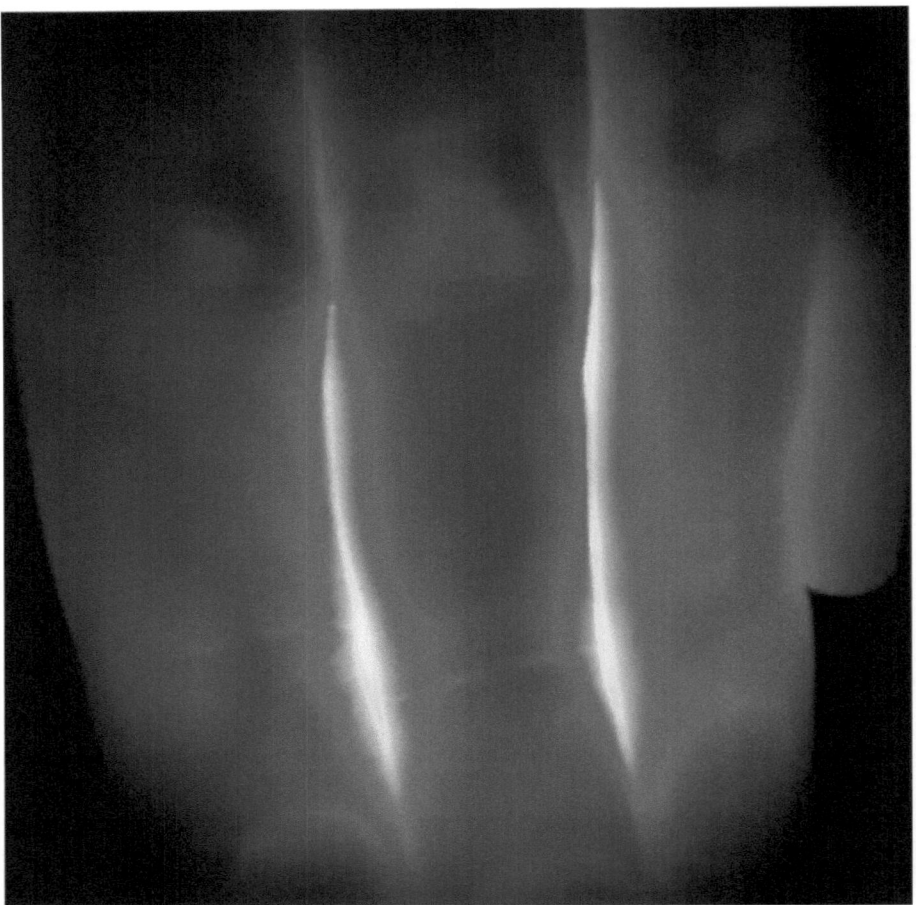

Figure 1. Color X-ray image of a hand.

A developed model of light resonance tomography (LRT) (Figure 2) consists of a gas discharge tube within which darkness is always present because this is related to the quantum medical theory of phosphorescent bodies. It is conceptually comparable to a flashlight device in which living beings lie with the purpose of recording deeper layers of these body cells. It also contains a gas discharge lamp, which is filled with xenon, for example. A high image quality is best checked with different gases, starting with the noble gases. A corresponding flash of light from all sides enables a digital camera

Kolek, Erik (2024). On the technological foundations of interstellar space travel. In: *Chronicles of Business Informatics Physics (CBIP)*. Volume 3, edition no. 1.0. ISBN: 9783759785237.

to take an image coincident with a Newtonian coordinate system (x, y and z axes). For this digital light resonance scanning, there must be absolute darkness in the gas discharge tube. Of course, the patient needs protective goggles for their eyes due to the extremely short but extremely strong light pulse. The LRT also has a computer with a voice console via which data can be spoken by the computer and recorded by quantum physicians, which makes the work of the quantum physicians much easier and saves working time while they view the currently recorded LRT single-stone images on a color image display. LRT scans usually take a maximum of five to fifteen minutes, also because the patient table can be quickly adjusted in height and sideways, which means that patients can be treated more efficiently on this type of ward.

Now, in the physical reality of nature, it is always the case that living beings that are exposed to too much photon radiation can also suffer corresponding damage to their health if the LRT is set incorrectly, but this is also already known from classic medical radiotherapy. The topic of light pulse intensity and light type is therefore discussed in principle below. An LRT should not be understood as a type of sunbed that can cause pathogenic changes to body cells both externally and internally, but as a type of sun flash device in which a vacuum can also be generated. Small mobile oxygen devices and larger stationary devices for supplying O_2 to patients via a tube already exist for life support. The vacuum can further increase the image accuracy of the digital image, as less flickering will be visible on the images. The latter phenomenon is known from astronomy, where images taken through telescopes flicker due to the atmosphere, resulting in blurred images. Accordingly, the Fourier series are designed more precisely and the detailing and summation options are therefore more advanced than with all other classic imaging devices such as X-ray, CT and MRT devices.

The light beam to be generated is extremely short but of extreme intensity, which requires a high voltage to be determined. The type of light, on the other hand, is determined according to the desired wave quantum gravity, which is used in quantum medicine today, i.e. between two opposite points within a coordinate system,

Kolek, Erik (2024). On the technological foundations of interstellar space travel. In: *Chronicles of Business Informatics Physics (CBIP)*. Volume 3, edition no. 1.0. ISBN: 9783759785237.

corresponding electromagnetic static fields are built up within the points in four dimensions dxi. The photon beam starting from P_1 to P_2 moves within this reference system and is drawn correspondingly short or long. This also makes it easier to penetrate body cells or living beings as a whole without having to rely on harmful conventional X-rays. Einstein rays are not so much harmful to health, but at most comparable to prolonged sunbathing, if not probably even harmless to health. As all the mirror points on the geometric surface lie next to each other inside a cylinder, this is a multiple gravitational or light field magnetic tomography in which the photon radiation is thought of diagonally. With the help of this diagonal approach, it is therefore possible to focus the photon radiation from all sides inside the cylinder around a single body cell point or to take images only locally at or from this one point. Targeted Einstein images can also be obtained by rotating the photon beams if their mirror points rotate around the recording point, whereby the Fourier series can be made more precise again and all other devices mentioned should be obsolete. Since only energy is required in a very short period of time, the radiation exposure to photons and fields in the LRT is also thought to be minimal for the patient.

Kolek, Erik (2024). On the technological foundations of interstellar space travel. In: *Chronicles of Business Informatics Physics (CBIP)*. Volume 3, edition no. 1.0. ISBN: 9783759785237.

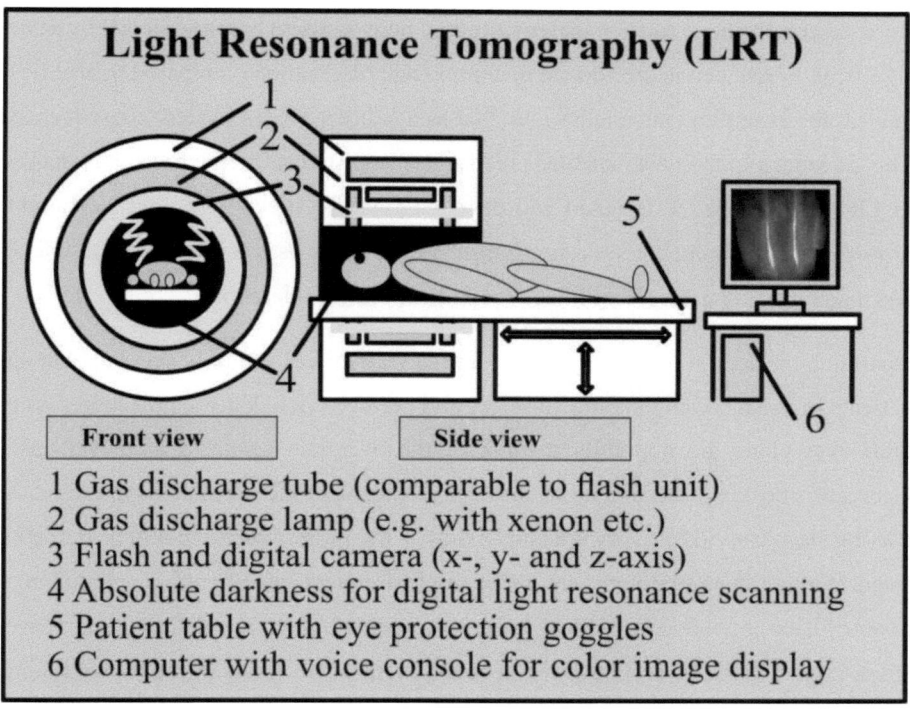

Figure 2. Light resonance tomography (LRT).

It should be noted at this point that quantum medicine is not subdivided into different medical specialties such as oncology, dermatology and neurology. No, this modern type of medicine looks at body cells, i.e. the living being as a whole, with the help of all medical specialties simultaneously, which is why only one single hospital ward per quantum physician and their patients makes sense from an organizational point of view. This makes quantum medical hospitals smaller in terms of space and time, yet with greater service capacities that can be planned, managed and controlled, but the arithmetic medical economy should be completely dispensed with and instead the focus should be on free humanism. Quantum medicine based on Euclidean geometry is therefore similar to advanced general medicine. Every patient only goes to a general quantum physician to get their pill or medicine. Aristotelian referrals to other classical medical specialties thus become superfluous. The LRT enables this type of hospital

Kolek, Erik (2024). On the technological foundations of interstellar space travel. In: *Chronicles of Business Informatics Physics (CBIP)*. Volume 3, edition no. 1.0. ISBN: 9783759785237.

ward described above (also in different spatial body classes) because the LRT can be significantly reduced in size due to quantum technologization (compared to an MRT device, for example) and enables a biobed as a mathematical-physical consequence. This advanced biobed also enables precise laser-assisted surgical procedures thanks to LRT. The biobed I designed and developed has joysticks with which quantum physicians can operate the precision surgical tools (such as laser and metal scalpels) and also display the necessary image magnifications for operations.

Operating theaters in which physicians, armed with knives and only standing, cut up patients in a manner comparable to steaks and perform such delicate procedures with their eyes alone are hopefully a thing of the twentieth century's medieval past. Generally speaking, the fatigue of these physicians and the hormonal symptoms of waking sleep caused by overwork and anxiety in the form of stress should be borne in mind. Waking sleep symptoms are, for example, frequent yawning, frequent blinking, poorer vision, poorer imagination, poorer logic due to attention deficits, aggressive stress-related behavior and, in extreme cases, even sleepwalking in the subconscious with open eyes or eyes partially covered by the nictitating membrane. That is why all quantum physicians know about the physical state of waking sleep in humans and rest much more than is the case in today's medicine by avoiding stress and thus producing less of the hormone of fear that makes them tired and with this knowledge can now also (better) consciously differentiate between the waking sleep state and the waking state. However, a waking sleep state in humans could also be generated by a more strongly summed gravitational field g_{ik} based on quantum gravity theory. In modern quantum medicine, all types of operations should therefore be much safer for the patient, because the biobed also enables a support information system for quantum physicians during the operation thanks to the computer, for example to display images of incision techniques again and even to have automated operations mostly performed routinely, such as appendectomy and suture knot techniques. Patients live long and in peace.

Kolek, Erik (2024). On the technological foundations of interstellar space travel. In: *Chronicles of Business Informatics Physics (CBIP)*. Volume 3, edition no. 1.0. ISBN: 9783759785237.

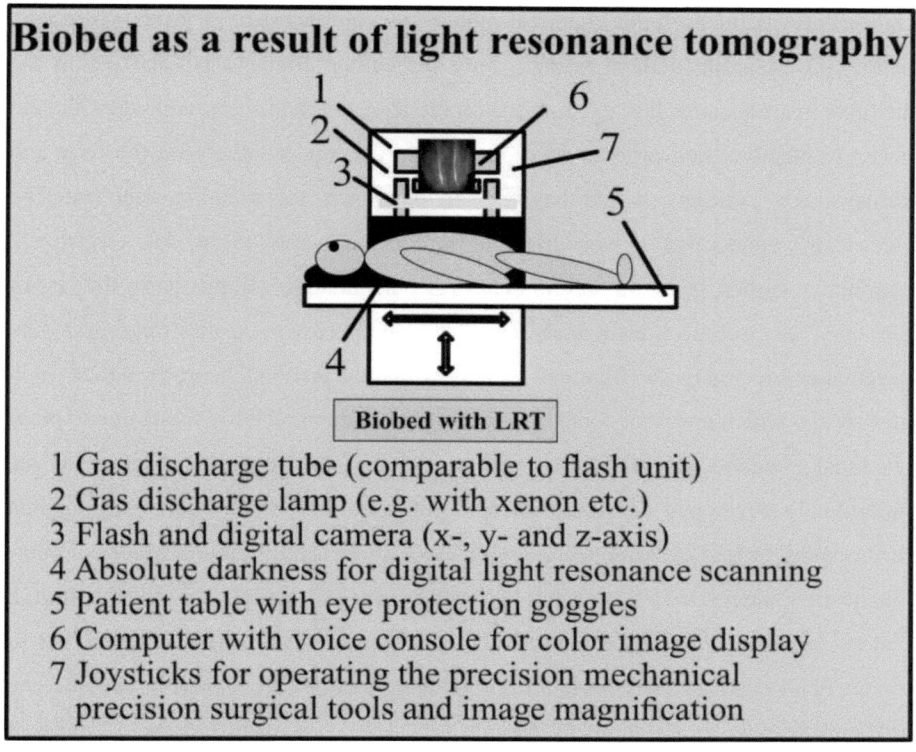

Biobed as a result of light resonance tomography

Biobed with LRT

1 Gas discharge tube (comparable to flash unit)
2 Gas discharge lamp (e.g. with xenon etc.)
3 Flash and digital camera (x-, y- and z-axis)
4 Absolute darkness for digital light resonance scanning
5 Patient table with eye protection goggles
6 Computer with voice console for color image display
7 Joysticks for operating the precision mechanical
 precision surgical tools and image magnification

Figure 3. Biobed as a result of light resonance tomography.

In keeping with the nature of living beings, the biobed can be used to perform various types of operations more safely for the patient, such as cardiac artificial heart surgery. The computer can be used to scan the patient's heart in the three-dimensional expression $\Omega = \int dx_1 dx_2 dx_3$. However, it should be noted that the patient's heart has become significantly larger due to cardiac wear and tear; the heart walls have therefore become pathologically thicker (as should be visible in the LRT Einstein image) and therefore also provide less pneumatic pressure. A corresponding compensation must therefore already be carried out during the scan of the patient's heart so that it can artificially regain its healthy natural geometry according to Euclid or span elasticity force by simply printing a coincident elastic artificial heart with the quantum medical 3D printer. Only two types of organic chemical elements are required for this, namely

Kolek, Erik (2024). On the technological foundations of interstellar space travel. In: *Chronicles of Business Informatics Physics (CBIP)*. Volume 3, edition no. 1.0. ISBN: 9783759785237.

carbon and silicon, each in a chemical rubber-like compound, to be used as filaments in the 3D quantum medical printer. Only the carbon filament is used within the artificial heart because it is in contact with the aqueous reddish metal hydroxide and no more metallic, non-organic compounds may get into it, otherwise the liver and kidneys, for example, would have to be examined for metallic deposits. The electrically conductive silicon filament is therefore applied to this electrically conductive carbon layer, so that an electrically conductive connection to the rest of the salty, electrolytic patient body can be created. However, this requires a tiny pacemaker covered by the filament, without which the artificial heart cannot be set in motion and which must not come into contact with other cell body tissue due to other chemical elements it contains, as this would certainly prevent rejection of the individually developed and fitted artificial heart by the patient's cell body. In this further miniaturized quantum technology device, there is only a rechargeable sodium-magnesium battery, which is probably non-toxic for cell bodies, instead of a lithium battery, which is definitely toxic for cell bodies, and which can deliver electrons to the artificial heart as desired via a thin-walled silicon wire wrapped in carbon. The artificial heart is also sutured to the coronary arteries with this particular type of quantum medicine wire. Due to the electrons, the artificial heart begins to beat in the patient's body, but the pacemaker can only start this artificial heart and keep it running for a limited period of time, but permanent operation of this battery device is not intended in quantum medicine. The reason for this is that this flexible elastic carbon-silicon heart works pneumatically with pressure and as soon as aqueous reddish metal hydroxide flows through it electrodynamically, electrostatic friction or charging occurs on the inner carbon, causing the artificial heart to contract in the same way as the tissue heart and expand again – as a result, even the battery of the artificial pacemaker is constantly recharged and hopefully never needs to be replaced. Conditionally, it should be able to beat and function independently after the electrostart impulse, whereby all types of cardiological diseases should no longer cause any discomfort to the patient thanks to quantum medicine, because the artificial

Kolek, Erik (2024). On the technological foundations of interstellar space travel. In: *Chronicles of Business Informatics Physics (CBIP)*. Volume 3, edition no. 1.0. ISBN: 9783759785237.

heart corresponds to a tissue heart and its natural physical function. The artificial heart described has the same geometric shape, mass, pneumatic movement function, electrical conductivity and is based on Maxwell's equations of electrodynamics, which also include electrostatics (Einstein, 1916), which I have transferred to the quantum medicine I have theorized and the associated quantum technologies, because living beings can also be considered analogous to electrolytic batteries and thus the same general laws of nature must be followed as should be the general covariant case for all bodies within our space-time continuum (Einstein, 1916).

References

Einstein, A. (1905). Über die von der molekularkinetischen Theorie der Wärme geforderte Bewegung von in ruhenden Flüssigkeiten suspendierten Teilchen; von A. Einstein. *Annalen der Physik 17(14)*, pp. 182–193.

Einstein, A (1905b). Über einen die Erzeugung und Verwandlung des Lichtes betreffenden heuristischen Gesichtspunkt. *Annalen der Physik* 322(6), pp. 132–148.

Einstein, A. (1906). Zur Theorie der Brownschen Bewegung; von A. Einstein. *Annalen der Physik 19(14)*, pp. 248–258.

Einstein, A. (1916). Die Grundlage der allgemeinen Relativitätstheorie. *Annalen der Physik 354(7)*, pp. 769–822.

Einstein, A. (2009). *Über die spezielle und die allgemeine Relativitätstheorie.* 24. edition, Springer.

Kolek, Erik (2024). On the technological foundations of interstellar space travel. In: *Chronicles of Business Informatics Physics (CBIP)*. Volume 3, edition no. 1.0. ISBN: 9783759785237.

Table of contents

Several medical quantum technologies are developed in this research article. These are light resonance tomography (LRT) and the biobed. These technologies represent advanced possibilities in quantum medicine. The idea of an artificial heart for humans has also been developed. The next step will be the development of corresponding prototypes for testing.

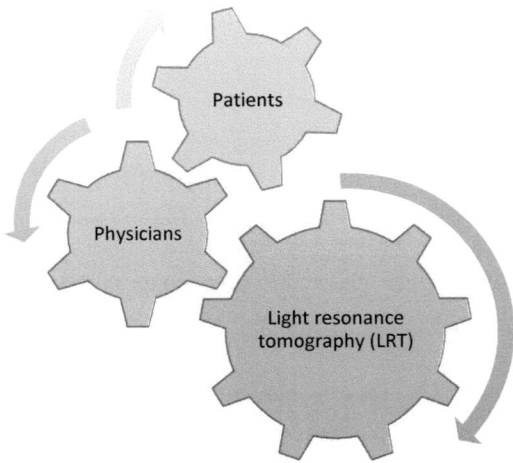

Figure 4. Table of contents illustrated by the importance of the theory of quantum medicine.

Kolek, Erik (2024). On the technological foundations of interstellar space travel. In: *Chronicles of Business Informatics Physics (CBIP)*. Volume 3, edition no. 1.0. ISBN: 9783759785237.

Scientific citation:

Kolek, Erik (2024). A theory on the development of a visor quantum technology as a quantum astrophysics-based possibility of a quantum medicine-based eye prosthesis designed primarily for humans. In: *On the technological foundations of interstellar space travel.* Chronicles of Business Informatics Physics (CBIP). Volume 3, edition no. 1.0.

Erik Kolek (2024)

A theory on the development of a visor quantum technology as a quantum astrophysics-based possibility of a quantum medicine-based eye prosthesis designed primarily for humans

Summary

Motivated by the theory of quantum gravity, which was developed by Erik Kolek as a theory of quantum relativity and thus represents an extension of Albert Einstein's special and general theory of relativity, this research paper today assumes a connection between quantum physics and astrophysics, or quantum astrophysics for short, and theoretical medicine, which leads as a consequence to theoretical quantum medicine and a theoretically possible quantum technology development. This particular development possibility is a visor quantum technology, i.e. a theoretically possible, i.e. possibly realizable, eye prosthesis, which is primarily intended for humans. A specially developed supersymmetric, relativistic model visualization approach supports the development of an understanding without the need for in-depth knowledge of business informatics physics. The development approach with regard to quantum technologies not only for quantum medicine was developed by an inventor who has research experience in the field of business informatics physics, but no experience in research and practice within (theoretical) medicine; this means that Erik Kolek is not a physician. As a researcher, he is therefore free from bias towards information and knowledge regarding the probably revolutionary theoretical quantum medicine as well as being truly neutral as an inventor towards the highly innovative theoretically conceivable development of quantum technologies such as the individual visor or universal visor. As a result, an artificial eye that can be adapted to personal needs as a prosthesis (individual visor) and, building on this, a more uniformly usable and therefore more cost-effective variation of this quantum technology (universal visor) will be designed step by step on the basis of quantum astrophysics. With the help of this theory, the quantum medicine-based development of this quantum sight technology is described by means of the quantum astrophysical possibilities, which should enable other inventors and researchers to successfully develop quantum technology, but unfortunately cannot or must not guarantee this success in general, because it is expressly not intended to raise false hopes among potential patients of physicians such as neurologists and ophthalmologists.

A theory on the development of a visor quantum technology as a quantum astrophysics-based possibility of a quantum medicine-based eye prosthesis designed primarily for humans

The results published by Erik Kolek (2024) on a quantum gravity theory already led to a very well-founded hypothesis regarding a theoretical, quantum physics-based medicine, which is to be derived today (Figure 1). The present research article is more related to the work of Albert Einstein (1905; 1916) and less to that of Stephen Hawking. According to the author, who has no research and practical experience in the field of (theoretical) medicine, their works, as a fundamental basis for quantum gravity theory, represent the groundbreaking physical explanation for this new revolutionary theoretical quantum medicine.

According to Einstein (1905, p. 174) and Kolek (2024), the following can be formulated: "The mass" (Einstein, 1905, p. 174) "and temperature" (Kolek, 2024) "of a" (Einstein, 1905, p. 174) "[...] very small" (Kolek, 2024) "body" (Einstein, 1905, p. 174) "are measures" (Kolek, 2024) "of its energy content, if the energy changes by L, the mass changes" (Einstein, 1905, p. 174) "and temperature" (Kolek, 2024) "in the same sense [...]" (Einstein, 1905, p. 174). Accordingly, according to Einstein (1905), Kolek (2024) obtains an optimized statement in the form of an increase in efficiency with regard to the energy $L = E = TMv^2$.

According to the first law of quantum gravity (1) by Erik Kolek (2024), all stars emit mass and radiation as light-heat photons that hit very large and very small bodies, which should influence and increase the gravity on these bodies. The very large and very small bodies that really exist in our ever faster expanding universe (continuum) would have to be pushed into the space-time of airless space (vacuum) by the heat of light from all the stars (Kolek, 2024). This would have to be caused by their own gravitation due to their own gravity of the existing elements as well as the added mass of the light heat photons when they hit the quasi-spherical surface of the very large and very small bodies (Kolek, 2024).

(1) Quantum gravity law $I = G \times [(T_1 M_1 T_2 M_2) / (r V_{Rel})^2] \times (+L/V^2 \times v^2/2)$ (Kolek, 2024)

According to the second law of quantum gravity (2) by Erik Kolek (2024), all black holes absorb mass and radiation as light heat photons that hit their event horizons, here very large and very small bodies are not reached by the light heat (photons), which would have to influence and reduce the gravity on these bodies. The very large and very small bodies would have to be pulled out of the space-time of airless space (vacuum) by the black holes absorbing the heat of light photons (Kolek, 2024). This is caused by their own gravity due to their assigned mass and the non-existent mass of the light heat photons, which now hit the event horizons of the black holes and no longer the very large and very small bodies (Kolek, 2024). Accordingly, when a black hole arrives in the space-time region of a very large as well as very small body, it should become dark on the respective body, even during the day, but this should be possible regardless of the size of a black hole (Kolek, 2024).

(2) Quantum gravity law $II = G \times [(T_1 M_1 T_2 M_2) / (r V_{Rel})^2] \times (-L/V^2 \times v^2/2)$ (Kolek, 2024)

Kolek, Erik (2024). On the technological foundations of interstellar space travel. In: *Chronicles of Business Informatics Physics (CBIP)*. Volume 3, edition no. 1.0. ISBN: 9783759785237.

According to the two laws of quantum gravity, it is likely that in an advanced quantum medicine a unified physical at least four-dimensional logic will appear possible as in quantum physics and astrophysics, abbreviated as quantum astrophysics (Kolek, 2024). The two laws of quantum gravity are modelled and visualized supersymmetrically, relativistically, so that their usefulness for quantum astrophysics and quantum medicine can be understood (Kolek, 2024). This model-based representation encompasses all previously explained physically conceivable facts (Kolek, 2024). The lower model representation represents an at least four-dimensional space-time domain of our ever faster expanding continuum (universe) (Kolek, 2024). This supersymmetric representation of relativity should make the two laws of quantum gravity (1) and (2) easier to understand even without in-depth mathematical knowledge. Both quantum gravity laws should be applicable in quantum astrophysics and quantum medicine, since they are likely to be in an interaction regarding the mass and radiation of stars or black holes with very large and very small bodies as well as in an interaction between the light or dark very small bodies (particles) with the light or dark very small bodies (photons). Due to the quantum gravity law I (1), mass particles that are not illuminated by light should not be able to be caught up by light-heat photons. Therefore, light-heat photons should also not be able to be caught up by dark very small bodies (particles) and, in general, the quantum spectrum probably appears dark based on the quantum gravity law II (2).

If we now look at the right-hand side of Figure 1 and think of the quantum gravity law I (1), we should notice that the human eye could behave like a particle, because bright photons hit our event horizon, which is our retina, and all living beings should be able to experience light. These living beings now understand the meaning of light by being able to perceive (observe) their surroundings, which is done with a certain number of eyes by seeing. The more light that reaches our event horizon (retina), the more the human pupil, an important part of the eye, generally changes or shrinks so that blindness does not occur. This should be a general protective mechanism provided by nature for all living creatures, which should have a corresponding variable eye

Kolek, Erik (2024). On the technological foundations of interstellar space travel. In: *Chronicles of Business Informatics Physics (CBIP)*. Volume 3, edition no. 1.0. ISBN: 9783759785237.

function by means of automatic calculation of the circumference by the diameter of the circle (π) in the brain. This should be a rather unconscious but nevertheless learned calculation process in the brain over which (human) living beings can exert little or no conscious influence. As a result, it should now be easy to understand that the quantum gravity law II (2) should lead to fewer light heat photons on the event horizon (retina), whereby (human) living beings should again experience less light. These creatures now understand the meaning of darkness by no longer being able to perceive (observe) their surroundings; here too, this is done with a certain number of eyes through vision, which is why some creatures such as us humans can see (observe) better or worse compared to cats, for example. The less light there is on our event horizon (retina), the more the human pupil would generally have to dilate as a part of the eye so that a higher combustion by light heat photons can occur, i.e. a theoretical possibility of seeing. It would therefore have to be a general opening mechanism provided by nature for all living beings, which would have to have a variable eye function by means of an automatic calculation of the circumference by the diameter of the circle (π) in the brain. This would have to be an unconscious but learned nerve process in the brain over which (human) creatures can exert little or no conscious influence, because otherwise they would be too busy with adaptations to their body or parts of it, such as the eye, and would therefore no longer be able to ensure their survival. This (human) eye function and other bodily functions should therefore have developed gradually through evolution and be passed on from generation to generation.

Kolek, Erik (2024). On the technological foundations of interstellar space travel. In: *Chronicles of Business Informatics Physics (CBIP)*. Volume 3, edition no. 1.0. ISBN: 9783759785237.

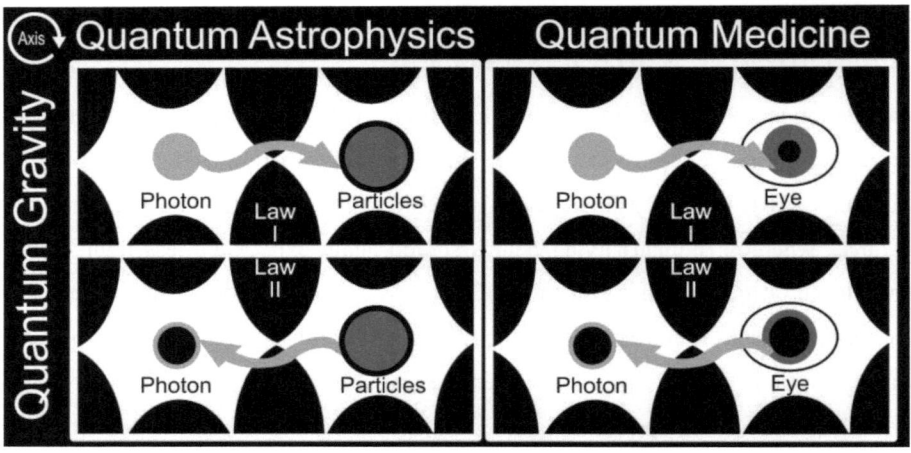

Figure 1. Quantum gravity as a theoretical link between quantum astrophysics and quantum medicine.

So if a high number of light heat photons hit the eye event horizon, a burn should occur on the retina of the eye, for example if the eye is imagined enlarged using the magnifying glass of quantum physics and without a pupil (Figure 2). If there were generally too long a time interval between a starting point and an end point due to an intense incidence of light heat, this should be easy to understand mentally as soon as an example of a switched-on space-time lamp, almost like a star, with a double superimposed square shape is thought of. However, this purely theoretical experiment is very dangerous and should therefore not really be carried out. Just as (human) living beings should never look directly into the sun (or an alien star) for too long a period of time, as a flood of light and heat would lead to blindness. Such a burn caused by too many light heat photons on the retina (eye event horizon) should be particularly easy for (human) creatures to observe or see in the dark, as just described. These (human) beings should therefore be able to see two additional red-colored squares in superposition with the remaining darker-looking thermal image according to their possible perception by mass and temperature; which should be formed on our eye retina (event horizon) with the help of different, non-Euclidean thermal areas. If this purely theoretical experiment is pursued further by reducing the number of light heat

Kolek, Erik (2024). On the technological foundations of interstellar space travel. In: *Chronicles of Business Informatics Physics (CBIP)*. Volume 3, edition no. 1.0. ISBN: 9783759785237.

photons located on the eye's event horizon, cooling should occur on the retina of the eye, i.e. if the eye is enlarged using the magnifying glass of quantum physics and without a pupil. If there were generally too short a time interval between a starting point and an end point due to a no longer intensive incidence of light-heat bodies, this should also be easy to understand mentally as soon as one thinks of a switched-off space-time lamp, almost like a black hole, with a double superimposed square shape. Such a cooling due to too few light heat photons on the retina (eye event horizon) should be particularly easy for (human) creatures to observe or see in bright light. These (human) beings should therefore now be able to observe two black-colored squares in addition in superposition with the remaining brighter-looking thermal image according to their conceivable observation by mass and temperature; which should be formed uniformly on our eye event horizon (retina) with the help of different, non-Euclidean thermal areas. This would then have to lead to not only sighted human beings being able to make corresponding observations (perceptions), but perhaps also not only blind human beings being able to learn to see again. At this point, it is important to point out that this paper is exclusively a theory, i.e. an assumption based on perceptible reality about existing reality, and that unfortunately no hope can or should be given for blind people, for example, even though this theoretical possibility of regaining sight by means of quantum medicine might really seem possible.

Kolek, Erik (2024). On the technological foundations of interstellar space travel. In: *Chronicles of Business Informatics Physics (CBIP)*. Volume 3, edition no. 1.0. ISBN: 9783759785237.

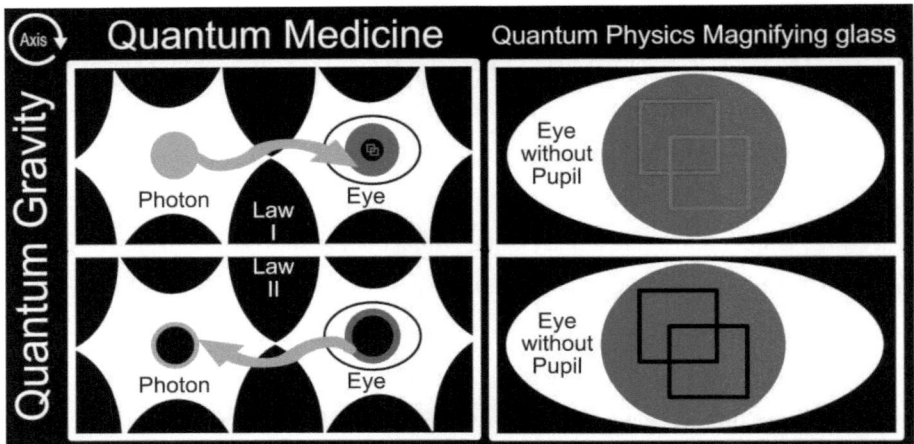

Figure 2. Quantum medicine of the human eye based on quantum gravitational physics under the quantum physics magnifying glass.

So far I have neglected the fact that the (human) eye event horizon is apparently not a two-dimensional circular surface (Figure 3). The (human) eye event horizon appears to me, if I now mentally consider the human eye under the quantum physics magnifying glass, to be more of a three-dimensional surface that does not quite correspond to a spherical geometry, i.e. more of a quasi-elliptical surface. Each eye should therefore have its own surface topology independent of the respective (human) living being, for example the lower eye looks somewhat irregular, although in general all eyes should function uniformly, even according to the special and general theory of relativity (Einstein, 1905; Einstein, 1916). From the outside, we can generally see the iris with its various structures and patterns based on quantum topology like recessed lines. Inside the eye event horizon, I imagine that these structures and patterns can be found again according to Albert Einstein's equivalence principle, i.e. simply put, they are repeated but concealed by a darkened, quasi-spherical type of light heat filter skin. It should now also be clear that the light heat photons do not hit or collide with a plane but with a curved space-time area in the (human) eye, which for some reason should be able to rise or fall according to the amount of light heat photons. This reason is introduced in the next paragraph. The color of the iris should

Kolek, Erik (2024). On the technological foundations of interstellar space travel. In: *Chronicles of Business Informatics Physics (CBIP)*. Volume 3, edition no. 1.0. ISBN: 9783759785237.

also have an influence on the visual ability of the respective (human) eye, the brighter it is, the better this creature can probably see, but unfortunately it can just as easily go blind. With a darker colored iris, it could really be possible to see better than other (human) creatures with a lighter colored iris, because they should have a worse theoretical possibility of going blind, at least not with a uniform rectilinear velocity movement even if a high amount of light heat photons should fall into the (human) eye. It could even be conceivable that there could be humans or other living beings who are really able to see the quantum world of physics, for example the small spherical but distorted air particles in our earth's atmosphere that bounce back and forth at different speeds. However, since we have been taught according to the standard model of physics that we cannot see the particles, we should really never be able to see them, since our brain could unconsciously convert this learned information accordingly; however, this is merely an as yet unevaluated assumption. So far, we have looked at the (human) eye under a quantum physics magnifying glass, but have not yet sufficiently taken into account its quantum topology, which should have a really decisive task with regard to the vision and blindness of living beings such as humans, but also animals.

Kolek, Erik (2024). On the technological foundations of interstellar space travel. In: *Chronicles of Business Informatics Physics (CBIP)*. Volume 3, edition no. 1.0. ISBN: 9783759785237.

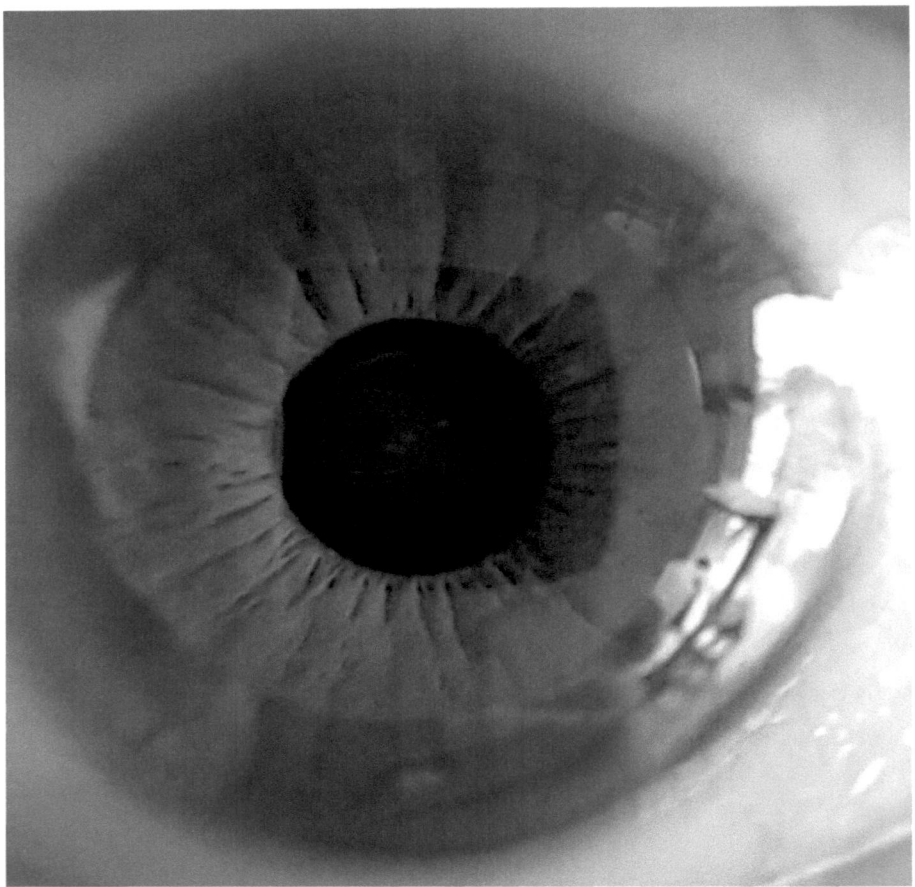

Figure 3. The human eye under the quantum physics magnifying glass.

If I now transfer the quantum topology just described to the quantum medicine of the (human) eye, I must determine the reason for the conceivable lifting or lowering within the iris (Figure 4). This reason for vision or blindness could be given by the fact that when light heat photons hit the quasi-elliptical eye surface, which is probably curved inwards in most visual organs, in large numbers, the body-warm iris mass TM increases in accordance with the added temperature mass TM of the light heat photons, and only in the space-time heat range (retina), This means that the eye event horizon (retina) decreases proportionally due to the reference body velocity v when the hotter photon mass strikes the colder body-warm cell mass. This collision of light-heat photons within our eye should cause us very slight, but nevertheless imperceptible pain, which not only we humans should be able to perceive or see via the nerve pathways to the brain as a visual field consisting of differently tempered and therefore differently painful point masses. This would also mean that our brain could only receive pain via the nerve pathways, i.e. even with pleasant, rather gentle touches; as a result, an existing pain scale would have to be adapted in line with this new quantum medicine-based neurology, with responsible physicians thinking about a more suitable reference system, which would also make anaesthesia much safer for life, for example, and all types of operations with fewer complications. If I imagine, although I am of course not a physician, purely mentally two superimposed Galilean coordinate systems K and K' now due to this necessity of a more innovative pain scale, then starting from a common point of origin of both reference systems, this should be positioned differently for each (human) living being, the negative sinking due to the burning of the light heat photons into the space-time heat range (eye event horizon) could be comprehensibly justified. However, this also means that, for example, the two overlapping square hotter body shapes K should lie lower than space-time regions K' on which colder radiating light heat photons strike, i.e. according to quantum medicine, the existing quantum topology of the (human) eye could probably influence the iris more strongly and therefore also lower it more strongly. In the same way, it should also be possible in the reverse eye model scenario related to the conceivable body universe (eye continuum) in the case of a cooling of the darker colored square space-time heat regions K, which should rise again proportionally as soon as the temperature mass TM of the iris and the light heat photons TM arriving in smaller numbers decreases, i.e. the space-time heat regions K' should rise again proportionally to the maximum to the point of origin correspondingly positive according to the reference body velocity v. Quantum topology is not the only way to illuminate areas of theory that still appear dark with knowledge, for example in the case of quantum

Kolek, Erik (2024). On the technological foundations of interstellar space travel. In: *Chronicles of Business Informatics Physics (CBIP)*. Volume 3, edition no. 1.0. ISBN: 9783759785237.

medicine, which can be derived on the basis of quantum gravity theory (Kolek, 2024); however, this is not a theory of experience, which should generally be built up inductively, i.e. derivatively, but on the contrary a deductive theory formation, which is intended to make light areas of knowledge that can now be theorized accessible to the inductive theory of experience via the truth; for example, for empirical research in quantum medicine.

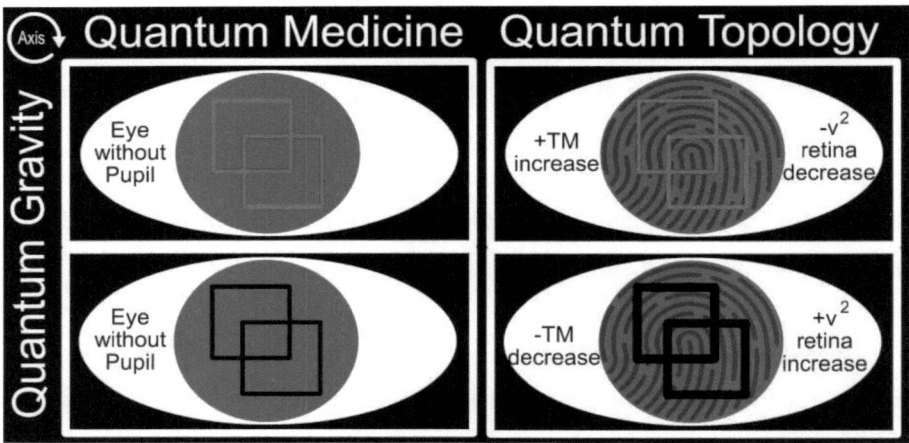

Figure 4. Quantum medicine based on quantum gravity physics and its quantum topology.

But since the (human) eye has a pupil whose light heat filter function should run unconsciously in the brain via π, because otherwise all (human) living beings would have to consciously control their pupil function as if they wanted to move another part of their body, the pupil must now be mentally assigned to the (human) eye again (Figure 5). The described eye model continuum (eye universe) should thus be exactly expanded and fully explained according to its function and be usable unchanged for a new, probably revolutionary quantum technology, which I have called visor quantum technology. The automated control of this sighting quantum technology could, for example, be carried out via the measuring physics by generally observing lumen changes, i.e. the described shift of the Galilean coordinate systems for the different space-time heat ranges of the artificial eye (K relative to K'), and in the event of an

Kolek, Erik (2024). On the technological foundations of interstellar space travel. In: *Chronicles of Business Informatics Physics (CBIP)*. Volume 3, edition no. 1.0. ISBN: 9783759785237.

increase in brightness, the pupil would have to be automatically controlled smaller accordingly so that no overexposed, white light heat image could arise in the (human) brain according to its neurology. In the case of increasing darkness, the measured numerical value for the lumen unit should therefore also decrease, so that the automated adjustment of the pupil would have to be larger, so that vision through quantum vision technology could theoretically again be a possibility for (human) living beings and possibly also an improvement in vision, for example for activities to be carried out in the dark, it would make sense to consider whether at least one artificial eye could be necessary and helpful. This could, for example, really be the physical case in the vacuum of our ever faster growing universe. Accordingly, visor quantum technology is a possible eye prosthesis that functions on the basis of a heat sensor further developed according to quantum topology with an integrated song function for sleeping and resting, because otherwise it would always be light and therefore this should otherwise have a disturbing effect. The content of the sighting quantum technology was fundamentally based on quantum medicine, which in turn is developed according to the theory of quantum gravity (Kolek, 2024), and therefore appears to be feasible in purely theoretical terms. Despite the existing theoretical work, a lot of work still needs to be put into the development of prototypes, which costs a lot of time and money, not only in enterprise cooperations, only in this way could a market-ready eye prosthesis be realized, should it really work, unfortunately the connection to the (human) body universe is still missing; this conceivable connection will be theorized below. At this point, I would like to reiterate the important note: I do not want to raise false hopes among all blind people and other living beings in the world, but merely point out and explain a theoretical possibility for a highly innovative visor quantum technology, so that their hopes could probably be fulfilled in the future; nothing more.

Kolek, Erik (2024). On the technological foundations of interstellar space travel. In: *Chronicles of Business Informatics Physics (CBIP)*. Volume 3, edition no. 1.0. ISBN: 9783759785237.

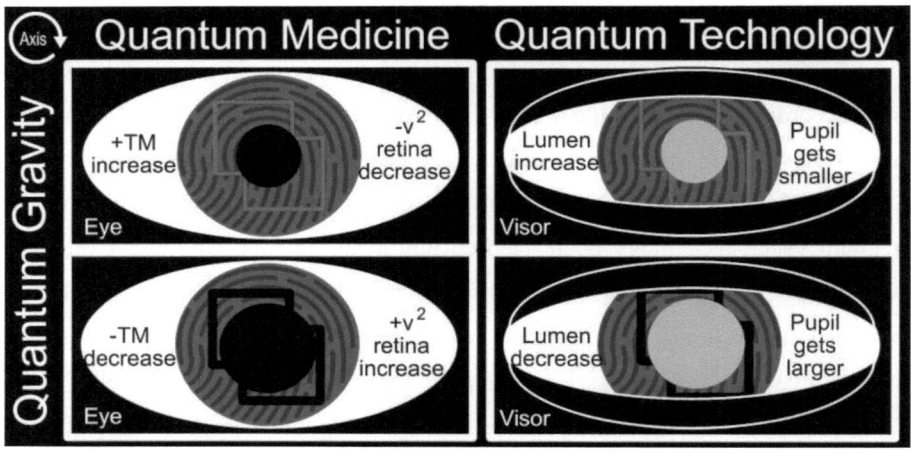

Figure 5. Quantum medicine based on quantum gravity physics as a basis for the development of a visor quantum technology (model 1 of 3).

In order to connect a quantum technology such as the visor with a (human) body universe, it is necessary to find out what living beings such as humans are actually made of, i.e. which elements of quantum chemistry exactly (Figure 6). It is generally known that humans should be made of carbon, which I cannot (yet) disprove, but it seems as if the human skin glitters under the quantum physics magnifying glass. I have also been able to discover this glittering on another person and different parts of the body, as can be seen in the picture below. In my opinion, the glitter is not dried sweat, which should also be found on human skin, but looks different, more like dried salt crystals, so I think we humans are possibly made of a carbon-silicon compound, especially our neuronal pathways could be made mostly of silicon and less of carbon. Silicon is generally known for its characteristic glitter, which I was apparently also able to see (observe) in humans; in particular, the edge of the image with measuring rods (test rods) is strikingly glittery and patterned; The surface texture therefore plays a less important role here, which mainly resembles non-Euclidean triangles and other geometric structures of a non-Euclidean nature such as quadrilaterals, which should also make it easier for physicians to check pathological changes in the human skin, for example whether a harmless birthmark is gradually developing into a black

Kolek, Erik (2024). On the technological foundations of interstellar space travel. In: *Chronicles of Business Informatics Physics (CBIP)*. Volume 3, edition no. 1.0. ISBN: 9783759785237.

melanoma. Perhaps this would allow physicians to further reduce the proportion of skin cancer in the world, and not just in human beings, by means of early detection thanks to the quantum medicine I have theorized. Quantum medicine should therefore really offer a wide range of new possibilities for the successful, proactive treatment of more than just humans; it is therefore a completely new research approach that I have in mind for human and veterinary medicine in particular.

Kolek, Erik (2024). On the technological foundations of interstellar space travel. In: *Chronicles of Business Informatics Physics (CBIP)*. Volume 3, edition no. 1.0. ISBN: 9783759785237.

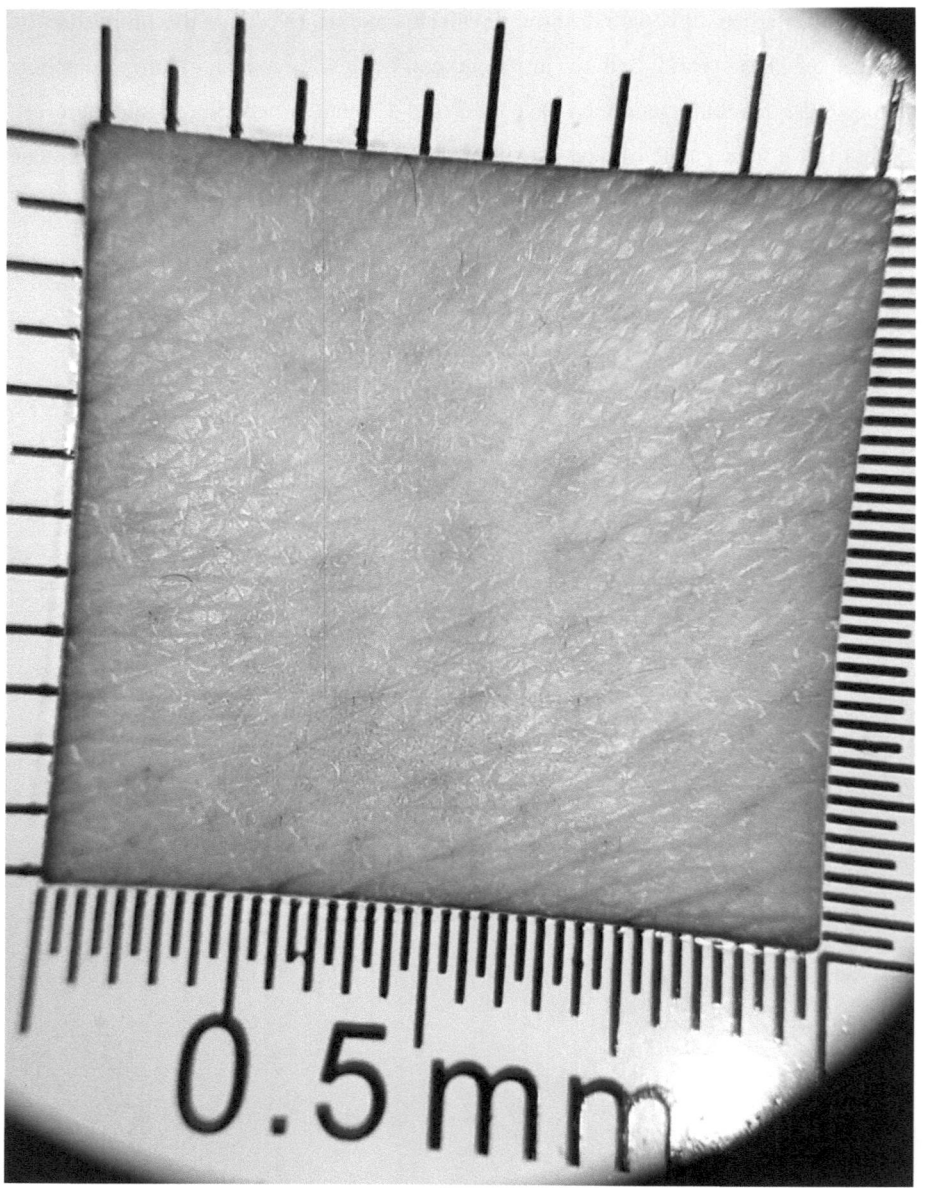

Figure 6. Human skin glistens under the quantum physics magnifying glass.

Kolek, Erik (2024). On the technological foundations of interstellar space travel. In: *Chronicles of Business Informatics Physics (CBIP)*. Volume 3, edition no. 1.0. ISBN: 9783759785237.

Finally, to connect a quantum technology such as the visor to a (human) body universe, I would like to propose a carbon-silicon nanotube for use in quantum medicine in practice (Figure 7). This flexible tube for connecting a quantum technology to the body therefore consists of an inner silicon conductor and an outer carbon heat blocker, which also forms the flexible sheath of the tube. The quantum topological, heat-sensor-based visor connected via the carbon-silicon nanotube could be anchored in the eye socket not only of humans via the optic disc with the optic nerve by, for example, placing a medical clamp, making a sclerotherapy, a simple deeper insertion or an incision with subsequent suturing; ultimately, a truly functioning connection could only be fully clarified by means of medical development of a prototype. However, the carbon-silicon nanotube I devised should generally represent the greatest potential to date for successfully connecting prostheses with the body via its nerve pathways, for example for eye prostheses but also for other (missing) body parts such as a forearm prosthesis with a hand that can be moved at will as usual. Quantum medicine could therefore form the basis for successful bionics, and not just for humans. Should this quantum technology really work, its function would have to be practiced by the patients accordingly, under certain circumstances movements would have to be relearned or colors reinterpreted, as these could be generated visually via the heat visor according to a human-machine interaction, if necessary one could even see (observe) only blue and red, but even this blue-red vision would be an excellent solution for a visor optics that is not only usable for humans. However, in the case of quantum technology, such as the visor, computationally intensive fine-tuning would always be necessary, because perhaps no or only a poor image arrives in the brain of the patient (test subject) and this despite all the quantum technological sophistication I have devised. This could possibly also be due to the frequency of the signal transmission, as well as other reasons that cannot yet be assessed, and although the carbon-silicon nanotube could really work; a misinterpretation of an apparently non-existent function would therefore be conceivable here. In the development of a medical prototype, attention should

Kolek, Erik (2024). On the technological foundations of interstellar space travel. In: *Chronicles of Business Informatics Physics (CBIP)*. Volume 3, edition no. 1.0. ISBN: 9783759785237.

definitely be paid to a possibly non-existent visual function in the interest of all potential patients; the conceived connection of the carbon-silicon nanotube might have to be optimized again by further reducing the diameter of the tube (possibly with a lower or no carbon content) and at the same time making the silicon conductor finer and even more conductive.

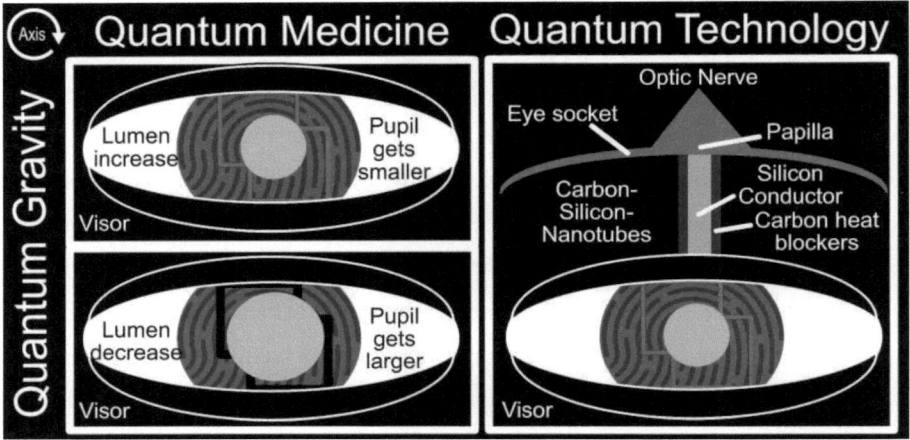

Figure 7. Quantum medicine based on quantum gravity physics as a basis for the development of a visor quantum technology (model 2 of 3).

Even more fantastic quantum technology variants of eye prostheses such as the visor are also conceivable (Figure 8). This is an alternative proposal for the development of a prototype with regard to visor quantum technology; here, as with the quantum technology generally modeled on the (human) eye, a night vision function could be integrated via the quantum topological heat sensor, which could be connected to the brain via two separate carbon-silicon nanotubes. The visor could therefore be divided into two sides A and B for the left and right eye technology connection to the respective optic nerve via the optic disc and follow a completely different design. This futuristic design could be thought of as a rectangle bent backwards over the (human) ears, similar to a pair of glasses, and instead of a pupil, functional slits with a uniform distance between them could be used. The clear advantage over an artificial eye visor

Kolek, Erik (2024). On the technological foundations of interstellar space travel. In: *Chronicles of Business Informatics Physics (CBIP)*. Volume 3, edition no. 1.0. ISBN: 9783759785237.

could be that it could be a cheaper, more standardized quantum medical vision device that would no longer have to be adapted to the respective eye socket and desired pupil shape like the deviating luxury visor variant; this could be the clear advantage of the universal visor as an innovative quantum technology variation. The visor should therefore have a thermal vision function and an individually adaptable, extended field of vision, which could be particularly useful at night or in the dark (in space travel); however, it could mainly be intended for use in dark, at least four-dimensional space-time areas of our ever faster expanding universe. Physicians would always have to weigh up the ethical reasons for its use in patients if, at some point, such universal or individual visors were to be used on space travelers as patients, even though they should still have one or more healthy eyes. So just because you could perhaps physically see better with a second artificial eye as a replacement for a (healthy) eye, for example at night, does not mean that it would be ethically sensible or justifiable for physicians. Other possible future fields of application for quantum technology variants such as the universal vision device or individual vision device could, of course, be machines not only in production, but also robots in general, which could perhaps even work for us in our households at some point, and possibly other living creatures such as animals; in veterinary medicine, for example, cats could therefore see more in the dark than people with healthy eyes, almost exactly as before. The power supply based on Maxwell's electrodynamics (Einstein, 1905) should also be mentioned as an existing quantum-technical limitation of quantum sight technology; the rechargeable lithium battery to be integrated should last for an estimated 12 to 24 hours at the current stage of development, i.e. a maximum of one day; here the universal sight still has an easily understandable advantage, because a much larger lithium accumulator could be used here. Finally, the following is still an important note: This paper is only a scientific but still futuristic theory, which is intended to describe a theoretical possibility for the development of a quantum technology such as the universal visor or individual visor, but in no way may or should knowingly

Kolek, Erik (2024). On the technological foundations of interstellar space travel. In: *Chronicles of Business Informatics Physics (CBIP)*. Volume 3, edition no. 1.0. ISBN: 9783759785237.

convey a guarantee or warranty with regard to the successful quantum technology development of a corresponding artificial eye prosthesis.

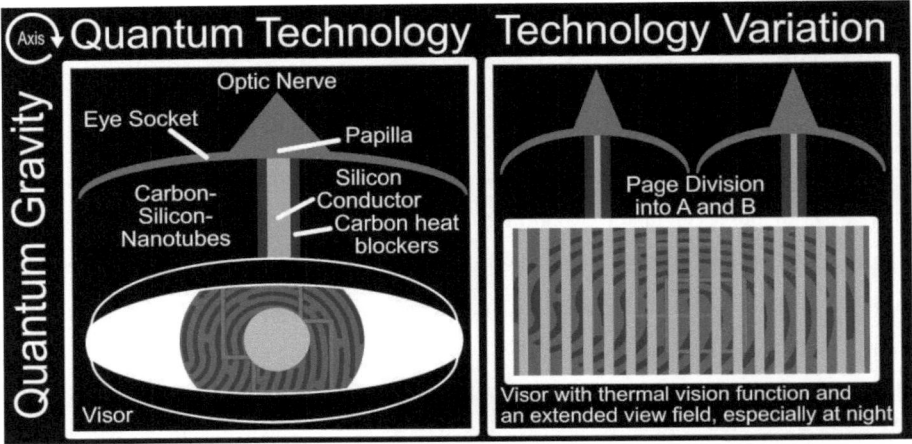

Figure 8. Quantum medicine based on quantum gravity physics as a basis for the development of a visor quantum technology (model 3 of 3).

References

Einstein, A. (1905). Ist die Trägheit eines Körpers von seinem Energiegehalt abhängig? *Annalen der Physik 18(13)*, pp. 639–641.

Einstein, A. (1916). Die Grundlage der allgemeinen Relativitätstheorie. *Annalen der Physik 354(7)*, pp. 769–822.

Kolek, E. (2024). Hängt die Trägheit eines sehr kleinen Körpers von seinem Energiegehalt ab?. In: *Über die physikalischen Grundlagen der interstellaren Raumfahrt.* Chroniken der Wirtschaftsinformatikphysik (CWIP). Band 2, Auflagen-Nr. 1.0.

Newton, I. (1687). *Philosophiae Naturalis Principia Mathematica.* 1. Auflage. Jussu Societatis Regiae ac typis Josephi Streater, London 1687 (http://cudl.lib.cam.ac.uk/view/PR-ADV-B-00039-00001/9 [visited on 08.05.2024]).

Kolek, Erik (2024). On the technological foundations of interstellar space travel. In: *Chronicles of Business Informatics Physics (CBIP)*. Volume 3, edition no. 1.0. ISBN: 9783759785237.

Table of contents

In this research article, a quantum technology is developed that is based on the theory of quantum gravity. It is a prosthetic eye for humans. It should be able to restore sight to blind people. The development of a prototype based on this theory will show whether this actually works or not.

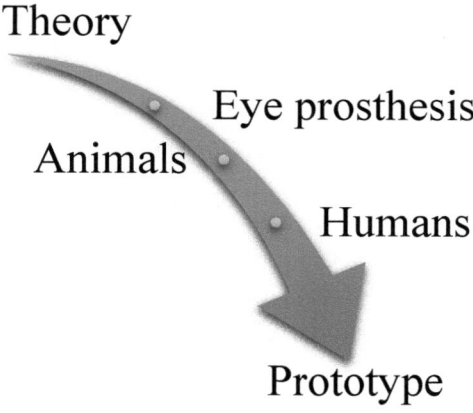

Figure 9. Table of contents illustrated by the importance of the theory of quantum technology.

Kolek, Erik (2024). On the technological foundations of interstellar space travel. In: *Chronicles of Business Informatics Physics (CBIP)*. Volume 3, edition no. 1.0. ISBN: 9783759785237.

Scientific citation:

Kolek, Erik (2024). The theoretical possibility of a quantum astrophysically based warp drive quantum technology with a maximum constant relative speed of light v with the maximum factor constant 10c. In: *On the technological foundations of interstellar space travel.* Chronicles of Business Informatics Physics (CBIP). Volume 3, edition no. 1.0.

Erik Kolek (2024)

The theoretical possibility of a quantum astrophysically based warp drive quantum technology with a maximum constant relative speed of light v with the maximum factor constant 10c

Summary

The present theory about a sovereign spaceship class starts with the first component, which consists of a theoretical possibility of a quantum astrophysically based warp drive quantum technology with a maximum constant light relative velocity v with the maximum factor constant of up to 10c. For this purpose, on the quantum astrophysical basis of Erik Kolek's quantum gravity theory, a more easily realizable down-to-earth warp drive quantum technology with the imaginary type number ZC-2063 with only a maximum constant relative speed of light v according to the somewhat lower maximum factor constant of up to 2c is first derived, which should now be possible to build as a phoenix-fast space rocket. However, since an advanced warp drive quantum technology should be a generation of quantum mass temperature, a matching general law of nature had to be derived using the example of Earth and Mars, for this purpose it is learned from a Mars photo from NASA that wind devils exist on both very large bodies, which are to be understood as a thought example of existing differences in atmospheric particle mass temperatures. These temperature-mass-rich findings are in turn used for the quantum technological further development of the warp drive with the type designation ZC-2063, in order to enable the more advanced

Kolek, Erik (2024). On the technological foundations of interstellar space travel. In: *Chronicles of Business Informatics Physics (CBIP)*. Volume 3, edition no. 1.0. ISBN: 9783759785237.

quantum technology variation of the warp drive, now labeled E-1701, by lifting itself out of the subspace-time range and surfing along on a quantum gravity wave. Quantum gravity surfing could be made possible by establishing new field equations of quantum gravity based on Albert Einstein astrophysics; the maximum energy impulse here should be similar to that of a stellar supernova and therefore the E-1701 warp drive should only be tested for the first time at the edge of our solar system with extreme caution.

The theoretical possibility of a quantum astrophysically based warp drive quantum technology with a maximum constant relative speed of light v with the maximum factor constant 10c

The results published by Erik Kolek (2024) on a quantum gravity theory have already led to an excellently substantiated theory regarding a theoretical, quantum technology-based warp drive, which is now to be derived (Figure 1). This research article is related to the work of Albert Einstein and Stephen Hawking. According to the author, who has no research and practical experience in the field of (theoretical) engineering, their works provide a very good basis for quantum gravity theory (Kolek, 2024), which has enabled the groundbreaking, quantum astrophysical development of this new revolutionary, theoretically possible warp drive quantum technology.

According to Albert Einstein (1905, p. 174) and Erik Kolek (2024), the following statement can be true: "The mass" (Einstein, 1905, p. 174) "and temperature" (Kolek, 2024) "of a" (Einstein, 1905, p. 174) "[...] very small" (Kolek, 2024) "body" (Einstein, 1905, p. 174) "are measures" (Kolek, 2024) "of its energy content, if the energy changes by L, the mass changes" (Einstein, 1905, p. 174) "and temperature" (Kolek, 2024) "in the same sense [...]" (Einstein, 1905, p. 174). Erik Kolek (2024) thus obtains an optimized statement according to Einstein (1905) in the form of an increase in efficiency with regard to the energy $L = E = TMv^2$.

Kolek, Erik (2024). On the technological foundations of interstellar space travel. In: *Chronicles of Business Informatics Physics (CBIP)*. Volume 3, edition no. 1.0. ISBN: 9783759785237.

According to the first law of quantum gravity (1) by Erik Kolek (2024), all stars emit mass and radiation as light heat photons that hit very large and very small bodies, which should influence and increase the gravity on these bodies. The very large and very small bodies that really exist in our ever faster expanding universe (continuum) should be pushed into the space-time of airless space (vacuum) by the light heat photons of all emitting stars (Kolek, 2024). This should be caused by their individual gravitation due to their respective gravity of the available matter as well as the added mass of the light heat photons in the collision with the quasi-spherical surface of the very large and very small bodies (Kolek, 2024).

(1) Quantum gravity law $I = G \times [(T_1 M_1 T_2 M_2) / (r V_{Rel})^2] \times (+L/V^2 \times v^2/2)$ (Kolek, 2024)

According to the second law of quantum gravity (2) by Erik Kolek (2024), all black holes absorb mass and radiation as light-heat photons, which should collide with their event horizons, here very large and very small bodies should not be reached by the light-heat photons that should affect as well as weaken the gravitational field of these bodies. The very large and very small bodies should be pulled out of the space-time of airless space (vacuum) by the light-heat photon absorbing black holes (Kolek, 2024). This could be caused by their individual gravitation due to their combined mass and the non-existent mass of the light heat photons, which should now collide with the event horizons of the black holes and could no longer hit the very large and very small bodies (Kolek, 2024). Accordingly, if a black hole were to arrive in the space-time region of a very large as well as very small body, it should become dark on the respective body regardless of the position of a planet such as the Earth in relation to its star such as the Sun, but this should also be possible regardless of the size of a black hole (Kolek, 2024).

(2) Quantum gravity law $II = G \times [(T_1 M_1 T_2 M_2) / (r V_{Rel})^2] \times (-L/V^2 \times v^2/2)$ (Kolek, 2024)

Kolek, Erik (2024). On the technological foundations of interstellar space travel. In: *Chronicles of Business Informatics Physics (CBIP)*. Volume 3, edition no. 1.0. ISBN: 9783759785237.

According to the two laws of quantum gravity, it can be assumed that for an advanced quantum technology such as warp drive, a unified physical logic of at least four dimensions as in quantum physics and astrophysics (quantum astrophysics) does not appear to be out of the question (Kolek, 2024). The two laws of quantum gravity are modelled supersymmetrically, relativistically and visualized so that their usefulness for quantum astrophysics and quantum technology becomes understandable (Kolek, 2024). This model-based visualization encompasses all previously explained physically conceivable realities (Kolek, 2024). The lower model visualization represents an at least four-dimensional space-time domain of our ever faster expanding universe (continuum) (Kolek, 2024). The depicted supersymmetric reality of relativity should make the two laws of quantum gravity (1) and (2) generally easier to understand even without in-depth knowledge of quantum astrophysics (Kolek, 2024). The two quantum gravity laws should be useful in quantum astrophysics and quantum technology development because they should be in an interaction regarding the mass and radiation of stars or black holes with very large and very small bodies as well as in an interaction between the light or dark very small bodies (mass particles) with the light or dark very small bodies (light heat photons) (Kolek, 2024). Due to the quantum gravity law I (1), matter particles should not be able to be caught up by light heat photons (Kolek, 2024). Therefore, light heat photons should also not be able to be caught up by dark very small bodies (matter particles) and, in general, the quantum spectrum should therefore appear dark, derived from quantum gravity law II (2) (Kolek, 2024).

So far I have omitted an important thought; I would now like to add this thought regarding a translational motion of a reference body K with respect to the surrounding space-time domain as Galilean reference system K'. On the left side of figure 1 you can see a quantum astrophysically considered, open system, in which, due to the quantum gravity law I (1), it is probably not true that the smallest particles of matter, which we call quanta, could actually be physically hit by bright light heat photons; these should always move away a very short test rod (measured distance) due to quantum gravity; similar to what the poles of a magnet do, where the same poles repel each other and different poles attract each other, except that here different quanta should tend to repel each other and only matching quanta could collide, at least this should probably really be the quantum astrophysically justified case in the vacuum of our growing universe and due to its mass quantum temperature. On the other hand, due to the quantum gravity law II (2) in the open system, it can be assumed that even matter particles could probably never really catch up with dark light heat photons.

In this approach, the issue of friction as a reason for heat development on the reference body K was deliberately neglected, because the friction of the reference body K (spaceship) on the reference system K' (space-time domain) only really arises as an interaction through its movement as soon as translational speeds close to or beyond the constant relative speed of light c are reached. At this point, all attentive classical physicists should now say to me critically that nothing can be faster than the constant relative speed of light c and that this is predetermined by the Lorentz transformation as a limitation for the translational speed of reference bodies. Unfortunately, this is not quite true, because Albert Einstein (1905) had used the Lorentz transformation for the derivation of his special theory of relativity and, in the case of his general theory of relativity (Einstein, 1916), had often and explicitly pointed out that his theory of relativity contains at least this special but also the general part of the theory and that even more aspects of the relativity model could be added as long as a deductive way of thinking about relativity and, in particular, about the constant speed of light c is applied. In his work on his field equations of gravitation, Albert Einstein (1915) pointed out that the general theory of relativity is valid even without the special theory of relativity. The individual components of the theory of relativity can also remain independent of each other, but these could also be adapted if necessary if new experiences (observations) are confirmed by experiments, or if they are placed in a different theoretical relativity context. According to my interpretation of the special and general theory of relativity (Einstein, 1905; Einstein, 1916), the law of the constant speed of light c only applies to light-heat bodies as reference bodies, but nowhere in Albert Einstein's works does it say that this could also apply to other quantum bodies such as an electron mass or temperature mass. That is why I am quite clearly of the opinion that science up to now, especially in classical (theoretical) physics, is simply subject to a misunderstanding here with regard to the formation of theory, which seems to me simply necessary for the derivation of the entire theory of relativity (special plus general theory of relativity), probably Albert Einstein assumed far too much knowledge with regard to his research method, so that much simply remained incomprehensible for most classical quantum physicists. That's why I don't quote any other authors from quantum physics, because they mainly work mathematically, sometimes even trying to work with multidimensional integrals, completely neglecting a sense-generating physics that needs to be presented as comprehensibly as possible with equations, texts, models and visualizations. Business informatics physics in particular provides a completely new understanding of this by means of supersymmetric, relativistic modelling and subsequent visualization. This is generally the advantage of an interdisciplinary research approach, which should also make scientific progress easier and therefore much faster.

The warp drive with the type number ZC-2063 I have conceived would most likely resemble a phoenix-fast spaceship rocket with the usual design or be equipped with at least two particle accelerators (warp nacelles) for directional motion control, which are located inside the cylinder or, better still, fixed or extendable to the side. This is a down-to-earth warp drive quantum technology that seems directly constructible and testable on the basis of this quantum astrophysical derivation, but I would suggest introducing the new modern model space travel, similar to model ships that should travel their routes in a closely spaced liquid or, in this case, in a widely dispersed gas, because this should save a lot of development costs and thus enable its successful quantum technological implementation with a small prototype. It is absolutely necessary to document the first flight of such a rocket warp spaceship well; preferably via online video link and, if possible, in real time, so that the whole world can experience it at that moment. The range of this rocket warp spaceship should

Kolek, Erik (2024). On the technological foundations of interstellar space travel. In: *Chronicles of Business Informatics Physics (CBIP)*. Volume 3, edition no. 1.0. ISBN: 9783759785237.

unfortunately still be inadequate, with it one could only travel very quickly to Mars or at the very most plan enough earth time to Proxima Centauri b, but I do not believe that a life-friendly world or even extraterrestrial intelligent life would await us there, at most a much too hot rocky planet, which, however, could perhaps even be used in the not too distant present as a thermodynamics-based power plant for a space station as a tourist attraction and stopover for interstellar space travel. A correspondingly planned and implemented particle accelerator, which should be thought of like a warp drive nacelle, should, since it still moves within the space-time range and does not yet rest on the same space-time range, not even on one of the many gravitational waves of different strengths, i.e. high gravitational waves, as in a space-time sea of gas quanta instead of short-spaced liquid quanta, the approximate single c to a maximum of twice c of the speed of light v should be achieved. At the smallest level of quantum astrophysics, liquid quanta should correspond exactly to gas quanta according to Albert Einstein's equivalence principle; it is highly unlikely that any difference at the smallest quantum level would be detectable by observation through vision, even by experimental physics. The light movement constant c is used here as a scale (test rod) that has been experimentally confirmed by humans for light and is therefore fixed in order to illustrate individual sections of speed, here up to 2c, but not in the sense of a speed limit as would perhaps have to be given on a space-time spaceship orbit, because there should be no general flight time limit here, so the speed v could really be possible. In summary, this could actually be physically dependent only on the friction of the reference body K on the reference system K', but this also means that this friction would then have to be reduced or even neutralized if mankind ever wanted to safely plan, implement and control a warp drive quantum technology with a higher reference body speed than 2c.

The warp drive ZC-2063 is a closed system for deceleration and acceleration using at least two or more particle accelerators in order not to cause a rotation around its own center in a circle, which could be symbolically designated for different classes of spaceships with different sized warp drive nacelles for different sized rocket

Kolek, Erik (2024). On the technological foundations of interstellar space travel. In: *Chronicles of Business Informatics Physics (CBIP)*. Volume 3, edition no. 1.0. ISBN: 9783759785237.

spaceships. The system I should be thought of as the quantum technological realization of the open system according to the law I of quantum gravity (1) and the quantum technological system II should be equal to the counteracted quantum astrophysical interaction as in the open system according to the law II of quantum gravity (2). Every warp nacelle would always have to have both quantum technological systems I and II ready for use, because otherwise it would be possible to accelerate but no longer be able to decelerate, for example, which would of course pose a problem with regard to free-flying very large and very small bodies. Imagine, for example, that the reference body K is flying at the directional speed v and encounters another reference body K", such as a moon, in the reference system K' without having installed a warp radar, which should function at least up to c by means of light sensors; what would then happen should now be generally clear to anyone who understands the term collision. Freely calculable scales (test rods) must be defined for the time coordinates t as well as for the space coordinates x, y, and z, because only in this way could warp navigation function successfully for spaceship route planning in accordance with the simultaneity between the starting point such as the Earth and a target point such as Proxima Centauri b. Warp navigation to simplify spaceship route planning could be a simultaneity planning that appears to be composable from different space-time model aspects, because a non-Euclidean spherical or perhaps even better a non-Euclidean quasi-spherical space-time geometry could enable an exact trajectory to be planned as a geodesic to be determined as necessary for a warp price. I consider uniform rectilinear translational motions of reference bodies K as described in the special theory of relativity (Einstein, 1905) to be completely impossible, not only because of the general theory of relativity (Einstein, 19016) but also because of the quantum theory of gravity (Kolek, 2024), in particular, both theories of gravity, i.e. Albert Einstein's astro relativity theory and my quantum relativity theory, lead me to think that no geometry according to Euclid should be applicable to warp spaceship navigation, which is why the spherical distances to be planned, or probably more precisely elliptical distances, should not be

Kolek, Erik (2024). On the technological foundations of interstellar space travel. In: *Chronicles of Business Informatics Physics (CBIP)*. Volume 3, edition no. 1.0. ISBN: 9783759785237.

uniform and also not rectilinear; This is also simply due to the Gauss coordinate systems applicable to them. Recalculations of course corrections would therefore always have to be carried out as quickly as possible, which would require a corresponding computer; this warp computer would still have to be developed for this purpose, as it should be able to calculate a very high bit bus speed v, or at least that of the constant speed of light c, because otherwise every use of energy diverted to the warp drive would conceivably result in a collision or, by inference, a serious catastrophe for a spaceship (reference body); This should apply to the physical reality of interstellar space travel and possibly less likely for excursions within our solar system, which we can also call the first space-time sector of our universe to be colonized by humans.

The described warp drive nacelles of the quantum technology ZC-2063 could simultaneously correspond to an impulse drive, which, however, would not be generally applicable on planets such as Earth, but only within a space-time sector, because here the respective atmosphere consisting of densely packed, fast-moving gas particles would make it difficult or even impossible to maintain the law of conservation of energy. In principle, a drive impulse could be freely selected depending on the speed of movement required or desired to achieve a moderate local target in the respective space-time sector, assuming that only the energy calculated for this purpose would have to be diverted to the warp drive quantum technology. As long as the reference body K (spaceship) would move in the reference system K' (space-time sector), since both coordinate systems could be in motion, there would probably be a very small but probably significant reduction in speed with increasing distance traveled, as long as the drive impulse v was calculated to be less than that of the constant light body velocity c. By now at the latest, it should be mentally possible to understand that, in terms of content, the business informatics physics revolves around a general type or theoretical possibility of star travel research.

Kolek, Erik (2024). On the technological foundations of interstellar space travel. In: *Chronicles of Business Informatics Physics (CBIP)*. Volume 3, edition no. 1.0. ISBN: 9783759785237.

It is generally important to build such a spaceship rocket with the warp drive ZC-2063 not curved, as it could otherwise be the physical model case in the universe according to the general theory of relativity (Einstein, 1916), but uniformly straight, as it should be according to the special theory of relativity (Einstein, 1905), because the spaceship rocket, previously conceived as a (curved) geodesic line and now to be considered as a straight line, should then automatically have a lower frictional resistance everywhere, regardless of its starting point, which should also significantly reduce the frictional temperature mass due to the generated velocity, which is at least as fast as light, with a maximum factor constant of up to 2c at the reference body K (spaceship) caused by the translational motion in the reference system K' (space-time domain). To create a better general understanding, my following completely freely imagined (fantasized) Mars travel story can be presented pictorially.

Albert Einstein, Stephen Hawking and Erik Kolek have traveled to Mars together because they are supposed to dig a horizontal hole there. None of them know what this hole will be used for in the end, but all three individuals, who should be able to think independently and not depend on each other, know the function of a spirit level and each of them naturally thought of a spirit level as well as the shovels when preparing their own Mars luggage. However, the observers on Earth would have to do the same, so they would also dig a hole as usual, which could later be used for a foundation, for example. Stephen Hawking could now say the following to Albert Einstein and Erik Kolek on the Martian sphere: "*Why don't you take a photo of how you use the spirit level and send it to the observers on Earth, but they should take the same photo of themselves using their spirit level on the Earth's sphere.*" Albert Einstein could only have thought of this at first and then decided to pronounce the following axiom directly to Stephen Hawking and Erik Kolek: "*If we compare these two photos, we could observe (see) two moments of the same geometric nature and probably realize that although both holes were planned the same and even measured with the same spirit level according to physics, they could still differ.*" Why could this be so in physical reality, of course, everyone wonders, even the observers on the earth.

Kolek, Erik (2024). On the technological foundations of interstellar space travel. In: *Chronicles of Business Informatics Physics (CBIP)*. Volume 3, edition no. 1.0. ISBN: 9783759785237.

All individuals could begin to look at the two snapshots and, as a collective common understanding, would perhaps realize that the angle below the spirit level on a sphere should always differ slightly according to the quasi-spherical but non-Euclidean geometry at each measured mass point due to the different gravity (here compared between Earth and Mars), especially because two excavated different spherical surfaces should be visible on the snapshots. Now Albert Einstein, Stephen Hawking and Erik Kolek understand that their dug hole surface could really look completely different, i.e. not appear horizontal at all, although they could have dug it with the same physical measuring method (spirit level) as the observers on the Earth's sphere; this should be even easier to understand if, as an example, a circle is generally considered, as soon as the rigidly placed, practically (almost) horizontal spirit level is applied to a circle as a scale (test rod) and the air body in the center can be seen, then a very slight angular deviation should always remain due to the circumference of the circle by the diameter of the circle (π), although physically it could actually look like this to people, especially because people might seem to believe this illusionarily due to the spirit level (measuring method) they have also invented; Of course, this circle should in reality also correspond to a non-Euclidean quasi-circle (elliptical shape) due to a quasi-spherical, non-Euclidean spherical geometry; this observation could only be made with an understanding of geometry that is consistent with the universe, for example with a Newtonian telescope that I also used for my (private) astronomy research for a while.

This could therefore be a new advanced architecture not only for spacecraft rockets, but also, as a brief outlook, a more innovative, uniformly straighter architecture, i.e. a more advanced architecture in general, for example for buildings such as houses, could now appear possible, not only on Earth but also on any other planet such as Mars. Since mankind would now be able to move (approximately) as fast as light or even a little faster than light, up to 2c as the maximum factor constant, through the space-time range K' of our solar system to the nearest star systems, this would be the beginning of the space-time age of Christopher Columbus for mankind, so to speak,

Kolek, Erik (2024). On the technological foundations of interstellar space travel. In: *Chronicles of Business Informatics Physics (CBIP)*. Volume 3, edition no. 1.0. ISBN: 9783759785237.

but we could perhaps appear on the radar of other civilizations through this faster-than-light movement; the Fermi paradox will be discussed in more detail later.

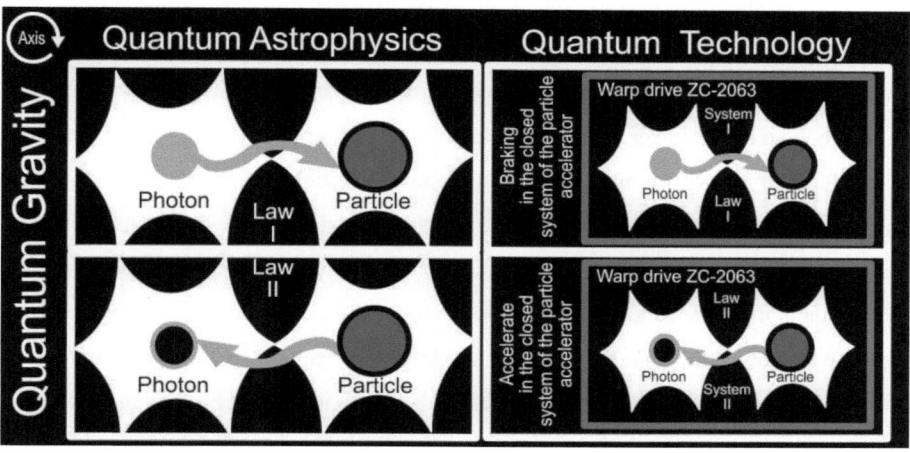

Figure 1. Quantum gravity as a theoretically possible link between quantum astrophysics and warp drive quantum technology with the maximum factor constant of up to 2c.

With regard to quantum gravity, I have so far only considered our Earth and its atmosphere in my explanations. Here we find oxygen particles, for example, which apparently, according to our experience, collide with light-heat photons in accordance with the first law of quantum gravity I (1), or rather are unlikely to actually exhibit this physical behavior, but rather should always experience a deflection shortly before an impact. It is therefore much more likely that a particle photon confusion should exist in reality, which we humans can perceive as a bright and yet transparent atmosphere, because if the particle photon acceleration were slower, we should be able to see (observe) it directly with the naked eye. It is probably the same when we humans see a dark, transparent atmosphere and ask ourselves why it is dark, a much better but theoretically possible answer to this physical question seems conceivable by the second law of quantum gravity II (2). If it is dark and no more bright light heat photons hit this hemisphere of the planet, such as the Earth, and thus can no longer transfer kinematic energy to other atmospheric particles, this does not mean that all

Kolek, Erik (2024). On the technological foundations of interstellar space travel. In: *Chronicles of Business Informatics Physics (CBIP)*. Volume 3, edition no. 1.0. ISBN: 9783759785237.

the available kinematic energy immediately disappears. No, this kinematic energy could be stored by the different atmospheric particles by constantly moving back and forth through quantum gravity, but could never actually touch or collide due to their different sizes, which therefore really exist as point masses of different sizes. In particular, this could be the physical case for dark anti-light heat photons, which should have a much lower charge of energy L, as they should perhaps have the same or more likely a lower mass, but should have a much lower temperature (mass fraction) and velocity. This should also be the reason why the dark anti-light heat photons have so far eluded the theory of experience and have not been discovered by experiments in measuring physics. At this point at the latest, we should have realized that the various atmospheric particles alone should not be the smallest quantum spheres but the smallest quantum ellipses. This elliptical shape of the quanta could be generated primarily by their own mutual gravitational energy, and this does not automatically mean that all elliptical particles could necessarily be moved uniformly in a straight line downwards by the gravity of the planet (Earth), although this is shown in Figure 2. The supersymmetric, relativistic model representation in Figure 2 with regard to quantum astrophysics based on quantum gravity physics can therefore not yet fully explain a particle weight difference, because for this the difference in terms of their temperature mass velocity (TMv^2) would still have to be better understood. I could have achieved this intellectual goal of understanding by carrying out a comparison between Earth and Mars to determine the difference in particle mass by imagining two different atmospheres in which, however, a particle wind should exist in both according to the general laws of nature.

For this derivation, it is enough for me personally to observe that there are wind devils on Earth and apparently, as can be seen in a NASA photo of Mars, wind devils also exist there; the text in this reference is not relevant and is therefore not quoted. In any case, I saw the wind devils in the Mars photo and so I took a mental journey from Earth to Mars together with Albert Einstein and Stephen Hawking. When we got there, we looked a lot bigger than an observer from Earth and we wondered together what

Kolek, Erik (2024). On the technological foundations of interstellar space travel. In: *Chronicles of Business Informatics Physics (CBIP)*. Volume 3, edition no. 1.0. ISBN: 9783759785237.

the reason could be, as we were all unable to explain it to each other as three observers on Mars? From the first glance at this Mars photo, it helped me to understand what a conceivable definition of mass means in terms of content, because understanding this better is the central aspect that generally needs to be learned if mankind really wants to be able to travel much faster than the speed of light with a reference body K (spaceship) through the reference system K' (space-time continuum) with only a maximum factor constant of up to 2c, if possible without frictional heat mass obstacles. An extremely well-designed hull armouring of the reference body K, similar to that of the solar satellites with electrodynamized carbon gold quantum compounds, could reduce this light speed barrier due to the probably arising, perhaps really braking frictional heat mass by a corresponding heat mass resistance, but probably never really prevent or even exclude it. Since it is generally known to me that the gravitation on Mars seems to be lower than on Earth, regardless of the exact numerical value determined by the measuring physics, it should now be possible for me to think with a research method based on Aristotle's way of thinking from Earth as a starting point in the universe and to judge with regard to Mars purely with thoughts, how the dark and bright mass particles as well as dark and bright light heat photons could possibly look like on Mars. In order to understand this, I also need the following conclusion: if there is less gravity on Mars than on Earth, then this observation behaves in the same way (analogously) for quantum gravity due to its given form and the general laws of nature; this should also be less on Mars than on Earth. Based on this Aristotelian way of thinking, I have probably learned correctly by means of deductive theorizing that quantum astrophysics could actually only differ in one physical aspect in general with regard to the planets, in this case Earth and Mars, namely in the geometry and size of the aforementioned mass particles and light heat photons. If I imagine, for example, that the world would only consist of Earth and Mars or only of the smallest mass quanta, i.e. the smallest fragments of Earth and Mars, then I immediately learn, based on Aristotle, that no spherical geometry could be applicable in the quantum spectrum, since correspondingly small point masses

Kolek, Erik (2024). On the technological foundations of interstellar space travel. In: *Chronicles of Business Informatics Physics (CBIP)*. Volume 3, edition no. 1.0. ISBN: 9783759785237.

could not be the smallest spheres. Such a spherical geometry of point temperature masses would probably correspond to a quasi-spherical geometry, i.e. rather an elliptical geometry. However, this elliptical geometry would not necessarily have to be oriented only towards the ground due to planetary gravity, as shown in Figure 2, but could also be conceivable in all directions, as quantum masses of different sizes would also have to exhibit different quantum gravity. To summarize, differences in temperature T measured as heat in Kelvin, mass M in the form of weight in kilograms and proportional reciprocal velocity v should really exist. This difference could only really enter the human mind as soon as the quanta are simply mentally superimposed, then a corresponding realization regarding the quantum difference in terms of temperature-mass velocity should become completely easy to understand. It is a pity that Aristotle did not yet know about the supersymmetric, relativistic modelling and visualization method I developed, otherwise science in general would already be further along with its progress today. Now observers from Earth would also have to understand, according to the fantasy-based, joint journey of thought to Mars by Albert Einstein, Stephen Hawking and Erik Kolek, why these three observers could have gained in body size and why the observers could have appeared smaller to these three; this phenomenon could probably be explained by the quantum gravity that exists there, which differs from that on Earth according to both laws; because this should act differently on each planet or very large and very small bodies.

Kolek, Erik (2024). On the technological foundations of interstellar space travel. In: *Chronicles of Business Informatics Physics (CBIP)*. Volume 3, edition no. 1.0. ISBN: 9783759785237.

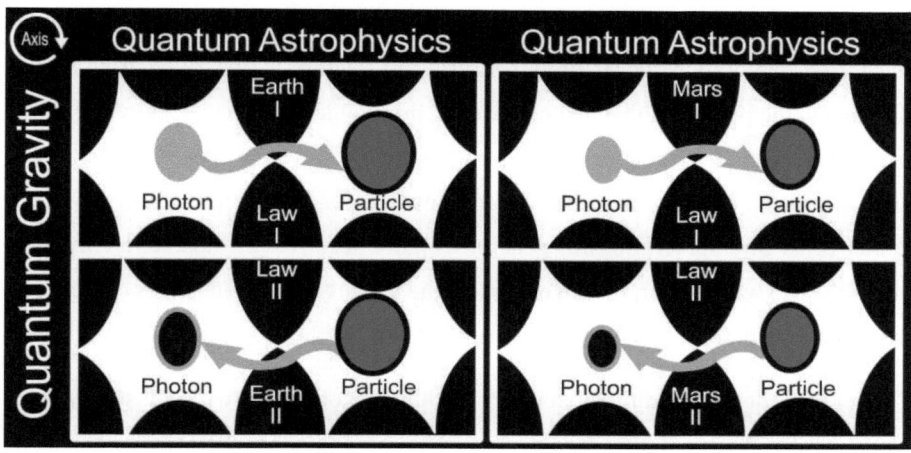

Figure 2. Quantum astrophysics based on quantum gravity physics in a comparison between Earth and Mars to determine the difference in particle temperature mass velocity.

The warp drive quantum technology with the type number ZC-2063 I came up with could once again be significantly improved in its maximum performance potential due to the realization of the differences in the speed of quantum temperature masses, so I had to come up with a new, more suitable type number; This number is now E-1701, where E stands for energy and because, as is generally known, in 1701 Isaac Newton published a general law of nature about the body temperature shift caused by air cooling, which is also known as Newton's law of temperature (Newton, 1701).

By developing this advanced warp drive quantum technology, the Fermi paradox, first posed by the nuclear physicist Enrico Fermi, may appear to have been finally resolved, as humans' understanding of quantum astrophysics and quantum technology development has not been sufficient and therefore we may not have been able to investigate our observations for possible encounters with non-human civilizations, our observations regarding possible encounters with non-human civilizations, so extraterrestrials could actually exist and they could even be waiting for us in other star systems, because interesting habitable planets could possibly also be highly sought after by other human or non-human civilizations. One approach to safe exploration

Kolek, Erik (2024). On the technological foundations of interstellar space travel. In: *Chronicles of Business Informatics Physics (CBIP)*. Volume 3, edition no. 1.0. ISBN: 9783759785237.

could be, for example, to disguise a space-time probe as an elongated asteroid (disguised exploration rocket probe) in order to venture a first flyby from a safe distance to the possibly inhabited new world; this would then be the beginning of the space-time age of Magellan, so to speak. A highly developed civilization whose technology could be far superior to ours would probably not hesitate, if it were a matter of its own survival or this generally highly negative, i.e. hostile, civilization, probably also waging war, to find out where the recently arrived intruders could come from and then perhaps pay them an unfriendly visit.

The reasons for an expedition to a super Earth in one of the more distant star systems could hardly be more diverse, starting with pure human curiosity, scarcity of resources, the development of a space economy and the increased necessity due to a lack of space on Earth caused by population growth. In this moment of Earth's rotational motion, it seems as if no super Earth with a temperature mass close to that of Earth is yet known to mankind, because only to such a super habitable planet would a faster-than-light war prize really be worthwhile, as humans would find it extremely difficult to adapt to changing gravitational conditions, at least until anti-gravitational quantum technology has been developed that could be implemented as an elevator without a rope in weightlessness for people or a space station with its own gravity for inhabitants. If the gravitational field of the planet were too high, we would collapse according to quantum gravity, and if the gravitational field were too low, we would feel stretched and have less stability in the skeleton of our bodies; both gravitational model situations would probably have pathological health consequences for humans. Other colonization issues still to be clarified, such as the special flora and fauna of the respective planet or microbes living there, would then certainly also play a role in human survival; at the very least, such tiny creatures would find it extremely difficult to survive free-flying in the planet's atmosphere, as they could be directly hit by the mass temperature quanta at a proportional speed that is at least as fast as light due to the likely constant quantum collision between (bright or dark) light heat photons and other (bright or dark) mass temperature quanta. Planetary adaptations (terraforming) would become generally relevant; these should start with the atmosphere. All in all, it therefore seems to be essential to launch and send out expeditions with sovereign spaceship classes occupied by crew members, which should always be equipped with at least two warp drive nacelles to improve the warp drag flow, which could cause a reference body K to flutter very quickly and thus make far-reaching course deviations appear possible. For the general improvement of this warp flow behavior, similar to the construction of bridges, in particular rope suspension bridges, two warp nacelles should always be used for more warp flight stability and less warp turbulence flutter, therefore the shape of the warp drive nacelles should resemble that of an inverted aircraft wing, This could further increase the adhesion of the reference body K (spaceship) on a gravitational wave emitted by very large and very small bodies from one space-time area to the next space-time area (reference system K' to K"). However, until today it could have been extremely difficult to travel to such a super-habitable Earth at a disproportionately high speed of light, but this should now finally change with the following train of thought as a far-reaching cork pop (limitation thought break).

If the sovereign spaceship class is really to be able to move faster than light with a maximum factor constant of up to 10c, like a single electron that has just been shot out of a supernova of a star, it would first have to lift itself out of the space-time area (reference system K') in which it is located at this infinitely short moment and permanently leave its own gravitational field as well as the gravitational fields of the other very large and very small bodies; otherwise there could be no silent acceleration of the quantum mass temperature caused only by quantum gravity, which cannot be heard without sound in a vacuum, with regard to the sovereign spaceship class but also with regard to the stellar supernova. The space-time region (reference system K') now located below the reference body K (spaceship) is therefore referred to for the first time as the subspace-time region (reference system K"), in which all very large bodies and associated very small bodies in particular should be located due to quantum gravity. In general, the space-time area (reference system K") located above the

Kolek, Erik (2024). On the technological foundations of interstellar space travel. In: *Chronicles of Business Informatics Physics (CBIP)*. Volume 3, edition no. 1.0. ISBN: 9783759785237.

subspace-time area (reference system K') could now also be conceptually defined as a class of a hyperspace-time area, but it should rather be a normal space-time area, which should much more likely allow a uniform rectilinear light body motion according to the special theory of relativity (Einstein, 1905), without having to consider the general theory of relativity (Einstein, 1916) and at the same time without being exposed to the constant risk of a sudden dimensional space-time range change due to quantum gravity (Kolek, 2024) as a possible danger, in particular a loop due to quantum gravity would otherwise be possible here, which could lead to the reference body K (spaceship) experiencing a translational motion shift and thus constantly changing its warp orbit course; catastrophic accidents could thus become the norm in the space-time travel of the 24th century of quantum astrophysics. Century of quantum astrophysics. With such faster-than-light translational motion speeds in the normal space-time range, a spaceship (reference body K) could really perform an unforeseen loop due to quantum gravity and, for example, have a collision with a very large body, just as this could generally be the fate of many an asteroid as a very small body; this could also be due to the fact that not only time is relative but also size is relative with regard to the bodies and their behavior. On the other hand, I would like to refer to a quantum space-time range (reference system K''') that is much smaller than the subspace-time range (reference system K'''), which is not relevant at this point, but could only be used for extreme transwarp speeds and other eleven-dimensional space-time tunnel projects. In order to maintain the clarity and size differences of the space-time areas described above, they are sorted according to their dimensions: at the bottom is the quantum space-time area (reference system K'''), in the middle is the subspace-time area (reference system K'') and at the very top is the normal space-time area (reference system K'); it should be noted that these at least four-dimensional cube areas, as Galilean coordinate systems, should themselves also be moving and not at rest. This space-time division was carried out according to the space-time curvature intensity to be expected there, just as Albert Einstein explained in his general theory of relativity (1916) after Riemann.

Kolek, Erik (2024). On the technological foundations of interstellar space travel. In: *Chronicles of Business Informatics Physics (CBIP)*. Volume 3, edition no. 1.0. ISBN: 9783759785237.

Accordingly, the light-speed motion of the sovereign spaceship class, viewed as a Galilean coordinate system, could first take place on the Z-axis before the translational motion would continue on the X-axis – almost without any frictional temperature mass generated by the speed against which the reference body K would otherwise have had to fly. Now it seems possible to use the producible energy for warp drive quantum technology much more efficiently than in the subspace-time range (reference system K"), from which it follows that the energy L approaches the Einstein-Hawking equation $E = TMv^2$ (Kolek, 2024) I have established, but will probably never be completely equal in space-time travel practice, because any losses, for example due to inefficient energy redistribution, starting with the warp fusion reactor via detours such as the anti-gravity stabilizers, which could prevent weightlessness in the sovereign spaceship, up to the warp drive, should always continue to be possible. If I start my travels from one point neighbor to another point neighbor, for example near the earth, the earth-space-time in which its specific gravitational field G-earth $(x, y, z, t)^2$ exists applies at the start, however, this gravitational field changes as soon as the reference body K approaches Mars, for example, which means that Earth space-time is always unequal to Mars space-time G-Mars $(x, y, z, t)^2$ but also unequal to the sidereal time G-Sun $(x, y, z, t)^2$ and so on, whereby the sidereal time should generally be a very good starting point for the determination of really existing planetary times. To determine the time on our clocks according to the causality that really exists in the universe, however, loop quantum gravity should be decisive as a consequence of quantum gravity, which will not be discussed further here.

Consequently, the basic features of the general (including the special) theory of relativity (Einstein, 1905; Einstein, 1916) can be further developed into the foundations of a general theory of super relativity for safe, convenient warp navigation. So if the faster-than-light class of spaceship flies from the Earth past the Moon to Mars within the space-time range of our Sun, this translational motion can be formulated using equation (3), which results directly in equation (4) for the general theory of super relativity, since the Minkowski space, as specified by the general

Kolek, Erik (2024). On the technological foundations of interstellar space travel. In: *Chronicles of Business Informatics Physics (CBIP)*. Volume 3, edition no. 1.0. ISBN: 9783759785237.

theory of relativity (Einstein, 1916), is also integrated. If the sovereign spaceship were now to fly on to Jupiter $G(x, y, z, t)^2$ at a maximum warp of 9.975c, a further G-$x(i)^2$ would become relevant in general terms, but this should now make a previously relevant G-$x(i)^2$ such as that of the Earth $G(x, y, z, t)^2$ appear irrelevant. G-reference systems would therefore always have to be automatically adjusted or updated in the at least light-fast warp computer as soon as a sovereign spaceship would generally travel into its vicinity, for which in turn corresponding at least light-fast warp sensors should appear necessary.

(3) G-Earth $(x, y, z, t)^2$, G-Moon $(x, y, z, t)^2$, G-Mars $(x, y, z, t)^2$, G-Space-Time $(x, y, z, t)^2$ = G'-Earth $(x, y, z, t)^2$, G'-Moon $(x, y, z, t)^2$, G'-Mars $(x, y, z, t)^2$, G'-Space-Time $(x, y, z, t)^2$

(4) G-$x(i)^2$, G-$x(i)^2$, G-$x(i)^2$, G-$x(i)^2$ = G'-$x(i)^2$, G'-$x(i)^2$, G'-$x(i)^2$, G'-$x(i)^2$

As just shown with the help of equations (3) and (4), such gravitational coordinates should be indispensable with regard to the at least four-dimensional space-time areas, which should be particularly important for interstellar travel, therefore gravitational coordinates should be a necessary condition for successful warp navigation, but may also appear useful for other quantum technologies such as direct mass temperature transport from A like Earth to B like Mars; for such long distances, however, very large amounts of energy would be necessary not only for signal amplification but also for signal transmission; a further prerequisite for the successful beaming of (perhaps also living) mass temperature bodies should be loop quantum gravity and an improved, thus more advanced radio shortwave telescope. I have now learned that gravity, including quantum gravity, must also move; I therefore call this Einsteinian gravitational quantum mechanics or Einsteinian warp quantum field theory, which contains the gravitational coordinates for the model representation of different space-time curvature intensities according to Riemann as in the general theory of relativity (Einstein, 1916) of the continuum (universe) and therefore also the Minkowski space.

Kolek, Erik (2024). On the technological foundations of interstellar space travel. In: *Chronicles of Business Informatics Physics (CBIP)*. Volume 3, edition no. 1.0. ISBN: 9783759785237.

In general, as soon as a reference body K (spaceship), starting from $G\text{-}x(1)^2$, would now approach $G\text{-}x(2)^2$ according to its (current) position through a subsequent approximation to $G\text{-}x(2)^2$, within which a gravitational field of lesser strength could act remotely, a positive or negative change in the energy to be applied or the necessary momentum should also take place according to the law of conservation of energy or momentum by Albert Einstein (1915), if the reference body K (spaceship) wants to be able to continue traveling at a constant relative speed of light of 9c, for example. This postulate of the super relativity theory according to Einstein's field equations of gravitation (Einstein, 1915) should also be comprehensible by the following mathematical equation (5), simplified according to its physical meaning, if one looks at the energy tensor of the respective gravitational field.

(5) $xt(i)^2 = \frac{1}{2} \times G\text{-}x(i)^2 - G\text{-}x(i)^2$ (each dependent on two reference bodies or reference systems) (Einstein, 1915)

However, if a different relation should become necessary with regard to the space-time range, it would have to be considered whether an oversized consideration of the respective space-time range might appear to be necessary. This would make particular physical sense for interstellar journeys of reference bodies K (spaceships), because the other reference systems not belonging to the reference system to be reached could also have a positive or negative effect on the warp trajectory curve $xt(i)^2$ according to Einstein's field equations of gravitation (Einstein, 1915). This new physical sense is added to the still simplified Einstein field equation (Einstein, 1915), in particular to make the summation sign used by Albert Einstein easy to understand.

(6) $xt(i)^2 = \frac{1}{2} \times G\text{-}x(i)^2 \times \sum G\text{-}x(i)^2 - \sum G\text{-}x(i)^2$ (The sum sign now indicates two interdependent reference systems) (Einstein, 1915)

Einstein's field equations of gravitation (Einstein, 1915) given here by (5) and (6) should also be applicable to the Kolek field equations of quantum gravity (Kolek, 2024), especially if two different reference bodies or reference systems (K_0 and K_1) accelerated according to the kinematics are generally considered. For the detailed description of the two quantum gravity laws I and II (Kolek, 2024) with regard to their mutual relative field strength, the following two new field laws therefore arise with regard to quantum gravity (7) and (8), which should also be applicable to subspace-time and other space-time curvature intensities.

(7) Quantum gravitational field law $I = \frac{1}{2} \times K_0 - K_1 = \frac{1}{2} \times G\text{-}x(i)^2 \times [(T_1M_1T_2M_2) / (rV_{Rel})^2] \times (+L/V^2 \times v^2/2) - G\text{-}x(i)^2 \times [(T_1M_1T_2M_2) / (rV_{Rel})^2] \times (+L/V^2 \times v^2/2)$

(8) Quantum gravitational field law $II = \frac{1}{2} \times K_0 + K_1 = \frac{1}{2} \times G\text{-}x(i)^2 \times [(T_1M_1T_2M_2) / (rV_{Rel})^2] \times (-L/V^2 \times v^2/2) + G\text{-}x(i)^2 \times [(T_1M_1T_2M_2) / (rV_{Rel})^2] \times (-L/V^2 \times v^2/2)$

These two quantum gravitational field laws can also be formulated in an even more general form for interstellar travel, analogous to the previous explanation, by observing the relative reference to the respective system as a sum sign to be integrated; ultimately, the translational motion should therefore be monitored and counteracted accordingly for every superluminal interstellar long-distance warp travel, i.e. if deviations of the warp trajectory curve occur, then the reference body K (spaceship) would simply have to be corrected accordingly so that the warp trajectory curve is maintained again.

(9) Quantum gravitational field law $I = \frac{1}{2} \times K_0 \times \sum K_1 - \sum K_2 = \frac{1}{2} \times G\text{-}x(i)^2 \times [(T_1M_1T_2M_2) / (rV_{Rel})^2] \times (+L/V^2 \times v^2/2) \times \sum G\text{-}x(i)^2 \times [(T_1M_1T_2M_2) / (rV_{Rel})^2] \times (+L/V^2 \times v^2/2) - \sum G\text{-}x(i)^2 \times [(T_1M_1T_2M_2) / (rV_{Rel})^2] \times (+L/V^2 \times v^2/2)$

(10) Quantum gravitational field law $II = \frac{1}{2} \times K_0 \times \sum K_1 + \sum K_2 = \frac{1}{2} \times G\text{-}x(i)^2 \times [(T_1M_1T_2M_2) / (rV_{Rel})^2] \times (-L/V^2 \times v^2/2) \times \sum G\text{-}x(i)^2 \times [(T_1M_1T_2M_2) / (rV_{Rel})^2] \times (-L/V^2 \times v^2/2) + \sum G\text{-}x(i)^2 \times [(T_1M_1T_2M_2) / (rV_{Rel})^2] \times (-L/V^2 \times v^2/2)$

As we have now certainly learned from the Kolek field equations of quantum gravity (7), (8), (9) and (10), a ride on a gravitational wave, which in general should rather resemble a quantum gravitational wave, should generally be a very high risk and thus a danger to be prevented for a deviation from the warp trajectory, because the reference body K (spaceship) could really disappear into nowhere, because it could

Kolek, Erik (2024). On the technological foundations of interstellar space travel. In: *Chronicles of Business Informatics Physics (CBIP)*. Volume 3, edition no. 1.0. ISBN: 9783759785237.

also be positively attracted by other galaxies and negatively attracted by black holes, which is why this model spaceship scenario within our thought model of the universe (continuum) should definitely be taken into account in the first warp test flights. Incidentally, the quantum gravitational field laws should also make it easier to predict the physical motion behavior of asteroids viewed as moving reference bodies within reference systems. As a quantum technological implementation of the quantum gravitational field laws with regard to the E-1701 warp drive, a correspondingly strong magnetic field would only have to be added to the warp nacelles, which are present at least twice on the sovereign spaceship, as a kind of turbo in the closed system of the particle accelerator, which should therefore make it possible to make the quantum temperature mass proportional to the speed much more predictable, manageable and controllable, i.e. the deceleration or acceleration in the closed system of the particle accelerator, but also rotations of reference bodies K (spaceships) should be made possible in a targeted manner. The E-1701 warp drive is therefore illustrated in Figure 3 according to the theoretically possible interaction based on the quantum gravity field laws between light or dark light heat photons with light or dark matter particles, which should make it easier to understand the two systems for deceleration and acceleration in the closed system of the particle accelerator, even without in-depth knowledge.

Finally, an extremely important note: physical experiments with regard to quantum gravity on the Earth (or other planets) to generate quantum gravitational waves with reference bodies could, under certain circumstances, really move them out of their perspective solar orbit, so this would be a real danger and not just a theoretically possible risk that must be taken into account in the interests of all humans, animals and plants. For example, it would be conceivable and if something seems generally conceivable in our universe, then it should also be realizable that the reference body K (spaceship) itself should also cause very strong quantum gravitational waves, which could then move away backwards during acceleration or forwards during deceleration higher like open semicircles and thus cause very strong uplift gravitational waves or downlift gravitational waves, which could also be able to lift other reference bodies

Kolek, Erik (2024). On the technological foundations of interstellar space travel. In: *Chronicles of Business Informatics Physics (CBIP)*. Volume 3, edition no. 1.0. ISBN: 9783759785237.

K' colliding with it – for example from the subspace-time range (reference system K'') – but which could just as theoretically be destroyed in the case of a very strong downforce gravitational wave, as this could behave in a similar way to an approaching uniformly wide, rectilinear black hole. To be on the safe side, applied research should first be carried out outside on the edge of our solar system with reference bodies K, K', etc., before flying faster than light with a maximum factor constant greater than 2c (in space-time travel practice approx. 1c) and up to 10c using reference bodies K, K', etc., as with sovereign spaceship classes equipped with at least two warp drive nacelles of type E-1701.

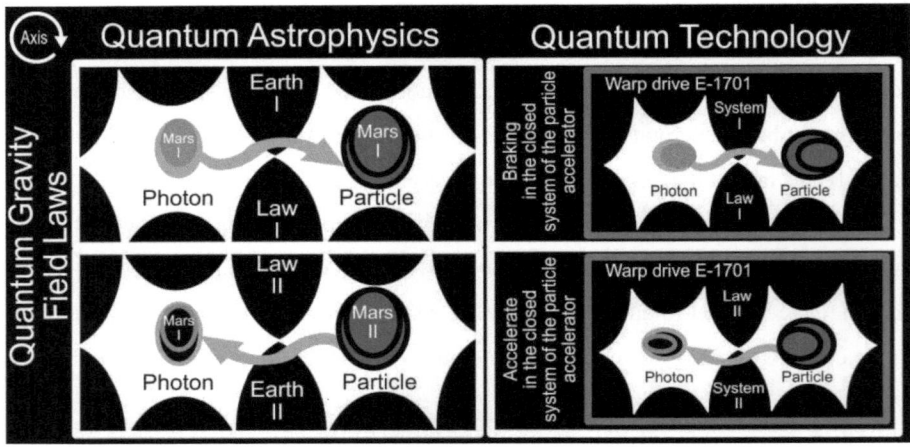

Figure 3. Quantum gravitational field laws as a theoretically possible link between quantum astrophysics and warp drive quantum technology with the maximum factor constant of up to 10c.

A brief addition: Just as with the rigid, practically (almost) horizontal spirit level, which only appears to be perpendicular, as a scale (test rod), there could also be a deviation of thought with regard to the surface geometry of the uniform hole to be dug on Mars if Albert Einstein, Stephen Hawking and Erik Kolek had each used their meter stick with distances on it that only appear to be equal as a scale (test rod) for measuring according to physics. Everyone on the earth should now agree that the same

Kolek, Erik (2024). On the technological foundations of interstellar space travel. In: *Chronicles of Business Informatics Physics (CBIP)*. Volume 3, edition no. 1.0. ISBN: 9783759785237.

observed phenomenon would also have led to different results according to the measuring physics in a space-time measurement with the help of practically (almost) fixed, identical clock faces with distances on them that only appear to be the same, i.e. time in general should not be the same as time on another body, which is why Albert Einstein only ever speaks of simultaneity (Einstein, 1905; Einstein, 1916) and not of time in general; probably there could even be no time at all but only body movements in our ever faster expanding universe (continuum).

As a brief outlook, it should also be mentioned that the foundation of the *anti-general theory of relativity* based on the work of Albert Einstein (1916) was derived by me as a further consequence for the field of quantum astrophysics, which in turn could theoretically make interesting quantum technological developments possible, such as a turbo-fast elevator possibly also up into the space-time range of our planet Earth, quantum gravity transformers to prevent weightlessness due to a lack of quantum mass temperature, especially conceivable for very large reference systems K', as well as inertial stabilizers in the centre of gravity intended for reference bodies K could theoretically become possible, so that very small bodies accelerated faster than light could also be maintained in a much larger reference body K. After the theorization of the basis of the *anti-general theory of relativity*, I would first like to publish an easy-to-understand book that once again explains all the basics based on Albert Einstein with regard to his general and special theory of relativity in an easy-to-understand way and introduces my *general-special theory of relativity* (light-body quantum mechanics) in a complementary and completing way. However, this could first be followed by a multidimensional television with a better, more precise, rotatable image representation, which could appear theoretically possible like an octagonal geometric not necessarily Euclidean shape in different very small and very large scale sizes (unit test scales) by means of quantum technology development; for this it seems to me today only necessary to better understand light rays and their relativistic symmetries not only on the basis of colors.

References

Einstein, A. (1905). Ist die Trägheit eines Körpers von seinem Energiegehalt abhängig? *Annalen der Physik 18(13)*, pp. 639–641.

Einstein, A. (1915). Die Feldgleichungen der Gravitation. *Königlich Preußische Akademie der Wissenschaften (Berlin)*. Sitzungsberichte: pp. 844–847.

Einstein, A. (1916). Die Grundlage der allgemeinen Relativitätstheorie. *Annalen der Physik 354(7)*, pp. 769–822.

Kolek, E. (2024). Hängt die Trägheit eines sehr kleinen Körpers von seinem Energiegehalt ab?. In: *Über die physikalischen Grundlagen der interstellaren Raumfahrt*. Chroniken der Wirtschaftsinformatik-Physik (CWIP), Band 2, Auflagen-Nr. 1.0.

Newton, I. (1687). *Philosophiae Naturalis Principia Mathematica*. 1. Auflage. Jussu Societatis Regiae ac typis Josephi Streater, London 1687 (http://cudl.lib.cam.ac.uk/view/PR-ADV-B-00039-00001/9 [visited on 08.05.2024]).

Newton, I. (1701). Scala graduum Caloris. Calorum Descriptiones & signa. *Philosophical Transactions*, pp. 824–829.

Marsfoto: https://mars.nasa.gov/resources/25235/curiosity-spots-a-dust-devil-in-the-hills/ [visited on 08.05.2024]).

Kolek, Erik (2024). On the technological foundations of interstellar space travel. In: *Chronicles of Business Informatics Physics (CBIP)*. Volume 3, edition no. 1.0. ISBN: 9783759785237.

Table of contents

In this research article, a quantum technology based on the theory of quantum gravitational fields is developed. Two laws of quantum gravitational fields are discovered. They are the basis for the development of the warp drive, which is a closed system of a particle accelerator. A magnetic field is added to enable more light speed.

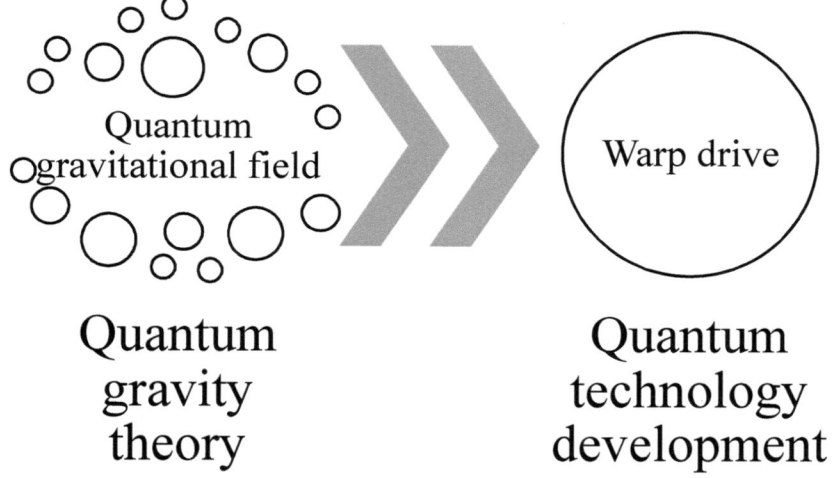

Figure 4. Table of contents illustrated by the importance of the theory of quantum gravity.

Kolek, Erik (2024). On the technological foundations of interstellar space travel. In: *Chronicles of Business Informatics Physics (CBIP)*. Volume 3, edition no. 1.0. ISBN: 9783759785237.

Third section: On the nuclear fusion of gases, an engineering method for technology development and the Erik-Kolek-Motor (E-KOMO)

Scientific citation:

Kolek, Erik (2024). On the theory of nuclear fusion of gases. In: *On the technological foundations of interstellar space travel.* Chronicles of Business Informatics Physics (CBIP). Volume 3, edition no. 1.0.

Erik Kolek (2024)

On the theory of nuclear fusion of gases

Summary

The aim of this research article is to describe the nuclear fusion of gases, including hydrogen to helium, from a theoretical point of view. A fundamental distinction must be made between hot and cold nuclear fusion of gases. Hot nuclear fusion of gases is the more dangerous of the two types of fusion. The latter is already being tested in physics and corresponding reactors are already in operation in experimental status, although not all questions of nuclear fusion of gases have yet been fully clarified. Black holes are also not supposed to be able to form during this hot nuclear fusion of gases, and if they do, then only very briefly and at the same moment these black holes would disintegrate again. I am of the opinion that once a black hole has been created, it also falls into the Earth's core. That's why hot nuclear fusion of gases should be treated with caution. With hot nuclear fusion, the aim of physicists is to imitate a sun on a small scale. At this point it is recommended never to go beyond a mass of 1 kg. It also makes little sense to pass plasma through magnetized rings while heating the mass.

Kolek, Erik (2024). On the technological foundations of interstellar space travel. In: *Chronicles of Business Informatics Physics (CBIP)*. Volume 3, edition no. 1.0. ISBN: 9783759785237.

On the theory of nuclear fusion of gases

This research paper deals with the nuclear fusion of gases because physicists have started to operate experimental reactors and because these have already caused plasma to ignite. This is all done without having understood the nuclear fusion of gases as a whole. Every hot plasma reaction poses a certain risk unless certain rules are observed when igniting the plasma. Stellarators and tokamaks are ring-shaped reactors for the generation of nuclear fusion of gases, which do not correspond to the model of the sun and would therefore have to remain in the experimental development stage. Current nuclear fusion reactors are not based on light and do not correspond to the structure of a star as a model. A sun is spherical or elliptical and the plasma must be shaped accordingly until the fusion reaction takes place.

A small sun rotates inside a functioning reactor for the generation of nuclear fusion of gases (Figure 1). The Aaron reactor, named after my cat, is an atomic reactor based on the neutrino reaction (AARON). Ideas from the Stellarator and Tokamak reactors were adopted to determine the magnetic field using technology. Here a small star rotates inside like a real star with a slightly lower temperature due to the lower mass of the star. The size of the star should initially be one millimeter in diameter when testing hot nuclear fusion. Later, as soon as more experience has been gained, larger spinning suns with more mass can be created as soon as there is certainty about their whereabouts when they go out.

In the external view, we have the transformer in shell design with the toroidal and external poloidal field coils. A poloidal magnetic field is based on a toroidal and resulting circular magnetic field. We therefore have a safer and stronger magnetic field. On the inside, we have pressure around the current plasma and laser technology to start the nuclear fusion of gases.

Kolek, Erik (2024). On the technological foundations of interstellar space travel. In: *Chronicles of Business Informatics Physics (CBIP)*. Volume 3, edition no. 1.0. ISBN: 9783759785237.

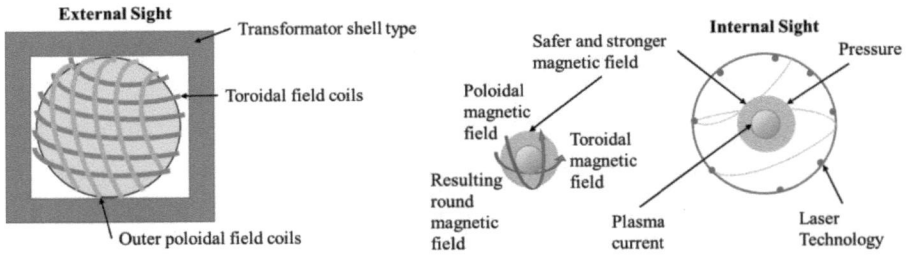

Figure 1. The reactor is called Aaron because of the neutrino reaction.

Creating a small spinning star should not be a problem in physics (Figure 2). A star is ignited using hydrogen as shown in the model. This is the principle of nuclear fusion of gases inside a star. A perfect spherical shape of the inner mirror with a large radius for safety ensures an undisturbed fusion power of gases.

But what happens as soon as the star is to be switched off by no longer supplying hydrogen? In my opinion, there are three possibilities.

Firstly, nothing happens, the star simply goes out.

Secondly, the star becomes glowing black and remains in place, emitting only heat until it can be removed as a metal ball after some time.

Thirdly, the star collapses and a small spinning black hole is also created due to the Earth's gravity. The lifetime of this black hole depends on its mass. The lower the mass of the black hole, the sooner the black hole disintegrates due to Hawking radiation [7, 8, 9]. It is possible that the small black hole will disintegrate on its own, so the magnetic field must be maintained at all times to keep the black hole in the center of the nuclear fusion reactor. The danger of a very small black hole is comparable to the danger of a very small sun inside a nuclear fusion reactor. It is important that the black hole is not fed any more matter, but must be kept in the magnetic field until it is completely destroyed by Hawking radiation [7, 8, 9]. This can take a few seconds or even longer until all the hairs of the black hole have also disappeared [9].

With regard to the very small rotating sun, Albert Einstein's general theory of relativity [1] and the Schwarzschild radius [2, 3] apply to the extinction of nuclear fusion of gases, which must also be taken into account, especially for the firing of hot nuclear fusion of gases. The Planck length [4] is also important in this context, as it represents the minimum size of a very small sun that is produced. A rule arises from the point of view that a very small rotating sun must be kept as small as possible in order to allow hot nuclear fusion of gases as safely as possible. The smaller the generated sun, the safer the hot nuclear fusion of gases. It is interesting to note that if hot nuclear fusion of gases works, gold and other precious metals should remain in the reactor as a waste product. It is also possible that other metals such as iron could be produced as waste, which would then have to be removed from the cold nuclear fusion reactor. Cleaning a nuclear fusion reactor is therefore a task that would have to be carried out from time to time or after each firing of the fusion furnace.

Inside the Aaron reactor there is a vacuum, atmospheric pressure, a magnetic field, the laser technology, the hydrogen in the center of the spherical structure to generate thermal energy. The outer shell of the reactor is made of gold, perhaps also of silver or aluminum. The curvature of light according to Alber Einstein's general theory of relativity [1] should be visible inside the reactor, also due to the magnetic field. The magnetic field corresponds to the gravitational field. Nuclear fusion of hydrogen to helium takes place right in the center of the sphere, and this in a very small star. The hot light photons should be able to stimulate this particle reaction. Gravity also plays a role in the heating of gases, because stars grow as long as matter is supplied. If no more mass is emitted, very small black holes could form, whose lifetimes need to be questioned and tested. For safety reasons, only very small stars should be created – especially at the beginning of the test series.

Kolek, Erik (2024). On the technological foundations of interstellar space travel. In: *Chronicles of Business Informatics Physics (CBIP)*. Volume 3, edition no. 1.0. ISBN: 9783759785237.

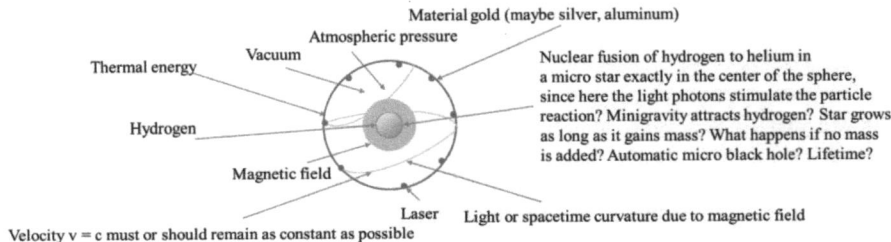

Figure 2. The Aaron reactor.

If I imagine a very large sun and a very small sun in a moving orbiting system, I realize that the radiation from the large sun should block the radiation from the smaller sun. At the quantum physics level, there should be differences in the size of the light photons that could suppress the radiation of the small sun. This is the photoelectric effect [10] between two suns. The light from the very small sun should be slower than the light c from the very large sun. This means that the wave-particle-heat trialism would have to be different. The radiation from the very small sun would have to have a lower speed than the light c and its photons would have to collide with the photons from the very large sun. Here too, energy would have to be generated according to the generally known equation $E = mc^2$ [6]. The photoelectric effect [10] of the momentum created by solar masses causes photons to merge. Nevertheless, some smaller bright photons should pass through the particle storm of the very large sun starting from the very small sun. A different light intensity therefore always means a different photon intensity. The energy of the inert small photon quanta should therefore differ depending on the photon speed and mass.

Cold nuclear fusion of gases is different from hot nuclear fusion of gases. The aim here is to successfully create so-called Bose-Einstein condensates [5]. These Bose-Einstein condensates [5] form the basis for the cold nuclear fusion of gases. It should be the safest form of nuclear fusion of gases. The creation of Bose-Einstein condensates [5] and almost simultaneous destruction of Bose-Einstein condensates [5] by heat produces energy in the known form $E = mc^2$ [6]. A reactor for this purpose has yet to be developed. There is already experience in the generation of Bose-Einstein condensates [5] in today's physics. Appropriate devices for carrying out the experiments should provide initial insights into the possible development of cold fusion reactors.

The easiest way to describe the function of cold nuclear fusion is to think of a very small drop of frozen water, which describes the hydrogen gas in a frozen state, the Bose-Einstein condensate [5]. This drop of very cold gaseous matter is now heated with pinpoint accuracy using laser technology. An explosive hot gas cloud and a very small spark are formed in a flash. This spark can be used to generate energy. This is a very efficient combustion of gases, as a rapid change of state from cold to hot would require more energy.

Similar results should occur as soon as very cold gas particles collide rapidly with other very cold gas particles. A very small cold gas fusion should take place, which should result in a very small cold gas cloud. This deflagration of cold gas matter should have a slightly hotter gas cloud. It could even happen that the cold gas cloud ignites and yet the cold Bose-Einstein condensate [5] remains behind. The collision of gas particles is therefore an alternative combustion option to laser technology, both of which could take place with pinpoint accuracy.

References

[1]   A. Einstein (1916). Die Grundlage der allgemeinen Relativitätstheorie. *Annalen der Physik 354(7)*, pp. 769–822.

[2]   K. Schwarzschild (1916). Zur Quantenhypothese *Sitzungsberichte der Königlich Preußischen Akademie der Wissenschaften zu Berlin* Januar – Juni, p. 548.

[3]   K. Schwarzschild (1916). Über das Gravitationsfeld eines Massenpunktes nach der Einsteinschen Theorie. *Sitzungsberichte der Königlich Preußischen Akademie der Wissenschaften zu Berlin*, Klasse für Mathematik, Physik, und Technik. p. 189.

Kolek, Erik (2024). On the technological foundations of interstellar space travel. In: *Chronicles of Business Informatics Physics (CBIP)*. Volume 3, edition no. 1.0. ISBN: 9783759785237.

[4]   M. Planck (1899). Über irreversible Strahlungsvorgänge. *Sitzungsberichte der Königlich Preußischen Akademie der Wissenschaften zu Berlin.* 5: pp. 440–480.

[5]   A. Einstein (1925). Quantentheorie des einatomigen idealen Gases – Zweite Abhandlung. *Sitzungsberichte der preußischen Akademie der Wissenschaften.* pp. 3–10.

[6]   A. Einstein (1905). Does the Inertia of a Body Depend upon its Energy Content? *Annalen der Physik 18(13)*, pp. 639–641.

[7]   S. W. Hawking (1974). Black hole explosions?. *Nature.* 248 (5443): pp. 30–31.

[8]   S. W. Hawking (1975). Particle creation by black holes. *Communications in Mathematical Physics.* 43 (3): pp. 199–220.

[9]   S. W. Hawking, M. J. Perry, and A. Strominger (2016). Soft Hair on Black Holes. *Phys. Rev. Lett.* 116, No. 23, 231301.

[10]   A. Einstein (1905). On a Heuristic Point of View Concerning the Production and Transformation of Light. *Annalen der Physik* 17, pp. 132–148.

Kolek, Erik (2024). On the technological foundations of interstellar space travel. In: *Chronicles of Business Informatics Physics (CBIP)*. Volume 3, edition no. 1.0. ISBN: 9783759785237.

Table of contents

In this research article, a nuclear fusion reactor is developed theoretically. Very small suns must be ignited in order to prevent very small black holes. Hydrogen is the basis for the ignition of these very small stars. The fusion to helium and other elements takes place inside the nuclear fusion reactor.

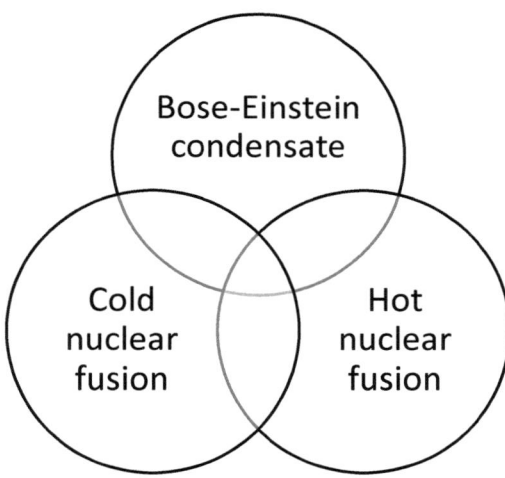

Figure 3. Table of contents illustrated by the importance of the theory of the photoelectric effect.

Scientific citation:

Kolek, Erik (2024). About an engineering method called Refit for the design of technological innovations. In: *On the technological foundations of interstellar space travel*. Chronicles of Business Informatics Physics (CBIP). Volume 3, edition no. 1.0.

Erik Kolek (2024)

About an engineering method called Refit for the design of technological innovations

Summary

This research paper deals with the engineering method called Refit for the design of technological innovations. The term refit is used to describe innovative inventions, patents and developments. There are large and small refits. Large targeted refits are so-called development projects, usually in interstellar space travel, and small refits, such as a single component, are created during the development process. The refit method is about understanding how innovations are created step by step. Refit is related to design thinking, which has its origins in business informatics and is also about the process-oriented development of innovations. The engineer who uses Refit is never one hundred percent satisfied with the innovation he or she has created. That is why he or she is always striving to optimize his or her refit. This is how new refits (innovations) are created. An attempt is also always made to produce a large refit; the small refits are rewarding developments on the way to the large refit. Nowadays it is common to work in a team, where there is always a chief engineer who is responsible for the refit project. The chief engineer decides how and when in the joint development process.

Kolek, Erik (2024). On the technological foundations of interstellar space travel. In: *Chronicles of Business Informatics Physics (CBIP)*. Volume 3, edition no. 1.0. ISBN: 9783759785237.

About an engineering method called Refit for the design of technological innovations

This research paper contains a summary of the most important ideas regarding an engineering method called Refit for the design of technological innovations. There is a certain relationship between the Refit method and innovation management and design thinking. The Refit method to be described has five phases, each of which has its own contribution to the overall development. First, these five project phases according to the Refit development method will be discussed. This is largely a theory of experience, as it is based on our own experience with the refit development method.

The refit development process will be illustrated to the reader using the example of the development of an electric motor (Figure 1). This refit process always begins with the definition phase, i.e. what is to be developed? In the example of an electric motor, this could be switchability, which should enable individual coils to be switched on or off so that they can produce energy due to induction while the electric motor is switched on. The definition phase is followed by the research phase. The aim of this research is to find out what has already been developed. Using the example of the electric motor, this could be its basic function, i.e. how such a motor works. The research phase is followed by the learning phase. It is learned how the refit can be developed. In the case of an electric motor, for example, this is done by determining the functions to be developed, such as its switchability. The learning phase is followed by the working phase. In this working phase, the refit is developed step by step until the desired functions are fully developed. In the case of the electric motor, this is the assembly of the electric motor with all new functions. The work phase is followed by the product phase. The product, or rather the refit, is fully developed and must be improved by definition, which means that the refit development process starts all over again. The finished electric motor is looked at again and given functions are developed further.

Kolek, Erik (2024). On the technological foundations of interstellar space travel. In: *Chronicles of Business Informatics Physics (CBIP)*. Volume 3, edition no. 1.0. ISBN: 9783759785237.

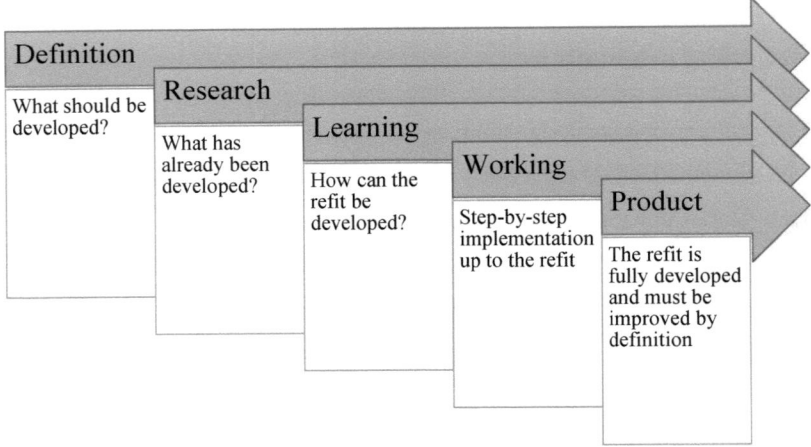

Figure 1. The refit development process.

This refit development method was primarily developed for interstellar space travel. This is where its greatest strength lies, because space projects are usually very large projects that can be difficult to keep track of. In these complex projects, engineers have the task of driving forward developments, which can sometimes also involve repairs. Some repairs also have the potential for a refit. Refit can be applied on a small or large scale, as it is primarily an innovative approach to the technology to be developed. It is therefore about the development of innovations on both a small and large scale.

Refits can also be created in a playful way by playing with copper, for example with the windings of an electric motor. The focus is then on having fun during development. When playing, some innovations are immediately apparent and can be implemented straight away. Engineers should often simply have fun during development, this is how refits are created in a playful way. Refits are fun for everyone involved.

During the research phase, it helps to look for existing patents, for example patents by Albert Einstein, Nikola Tesla and Thomas Alva Edison, because a lot can be learned from these patents. These developments were partly produced experimentally and partly only in thought, which means that small and large refits can also be created with imagination. Imagination is therefore more important than knowledge, as Albert Einstein said, even in the refit development process. It is important to understand that the individual refit development steps are flexible, which means that you can go one step forward or one step back at any time. For example, patents found can be used to go back to the definition phase in order to better describe the refit to be created. The patents found also create a deeper learning phase and the how in the refit development process will become clearer. With the implementation in mind, the work phase can be approached efficiently. The work phase is about building and constructing small and large refits. It is often the case that during the development of large refits, small refits are sometimes created unintentionally. In the product phase, the refit is completed and optimized directly. These optimization processes are characteristic of the refit development process steps. Any optimizations found are directly integrated back into the (new) definition phase.

Refit can also be applied to theoretical work, for example in scientific work, where the aim is to improve a technical paper or theory (Figure 1). A fit is a development and the refit is the optimization of this development. A fit is always followed by a refit, which becomes a fit, which in turn becomes a refit and so on. A refit development process is therefore always an optimization process from fit to refit to fit to refit and so on. In this context, an old refit can also be called a retrofit. The refit development process can also begin with the definition phase when improving a theory. For example, a theory by Albert Einstein is to be improved, which could be the special theory of relativity [1]. In the research phase, the original article by Albert Einstein is found [1]. In the learning phase, the article is read and understood and ideas emerge as to how this special theory of relativity [1] could be improved. In the working phase, the focus is then on writing these new ideas, for example with regard to quantum gravity; the refit is created step by step. The finished technical article as a refit of the special theory of relativity [1] then represents the product, which is also directly a fit

Kolek, Erik (2024). On the technological foundations of interstellar space travel. In: *Chronicles of Business Informatics Physics (CBIP)*. Volume 3, edition no. 1.0. ISBN: 9783759785237.

and can be further developed into a refit by definition. It is a living development process that leads to the refit. This almost endless development process will become slower and slower over time, because designing a refit always means seeing or recognizing the innovation that is to be implemented.

Creating a refit is fun and results in the creation of a new technology (Figure 2). The engineer is aware of his performance because he has internalized the refit development method over time. Experience therefore plays a major role in the refit development method. Experience with refit is specifically generated in each phase of the development process. The basics are created in the definition phase. In the research phase, existing knowledge is used. In the learning phase, new experiences are generated through social learning and understanding of technological contexts. In the work phase, the experience is created as the refit gradually takes shape. The product is then the completed refit and the experience gained, which is continuously optimized. A refit of the definition phase takes place. A further research phase and learning phase. An improved work phase and finally the engineer returns to the product, the refit.

In interstellar space travel, experience shows that there are complex development projects that need to be successfully completed. The refit development method is now available to the engineer for this purpose, and teams in particular benefit from collaborative development work. The chief engineer bears responsibility during collaborative development work. The spaceship is fully planned during the definition phase. Then, in the research phase, research is carried out into existing technologies that are suitable for use during the subsequent construction of the spaceship. In the learning phase, you learn how to assemble the spaceship step by step. The spaceship is created in the collaborative working phase through the cooperation of all engineers involved. The finished spaceship is to be optimized again as a product, because Refit always starts from the beginning with the definition phase.

Kolek, Erik (2024). On the technological foundations of interstellar space travel. In: *Chronicles of Business Informatics Physics (CBIP)*. Volume 3, edition no. 1.0. ISBN: 9783759785237.

The special thing about Refit is its simplicity, because without Refit there were only creativity techniques that were difficult to implement, such as brainstorming, creative writing, mind mapping, map queries and so on. This list does not claim to be exhaustive, it should only show that there are creative possibilities in collaboration, but no single development method like Refit. Design Thinking is related to Refit, as both deal with the optimization of solutions in terms of content. However, Refit is a far more advanced development method for designing innovative technologies. Design Thinking is therefore considered outdated and is not suitable for the development of a spaceship. Refit, on the other hand, is suitable for the development and construction of modern spaceships as we know them from the well-known series Star Trek. We could start with the Enterprise NX-01, because this spaceship represents the state of the art of technological developments today.

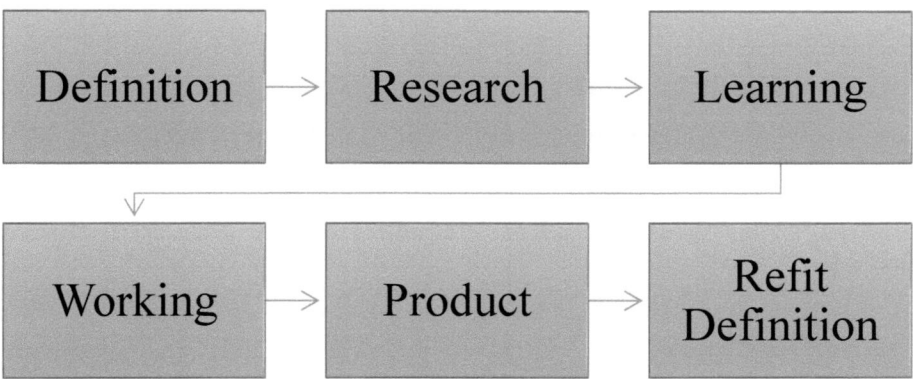

Figure 2. The refit development process steps.

References

[1]   A. Einstein (1905). Does the Inertia of a Body Depend upon its Energy Content? *Annalen der Physik 18(13)*, pp. 639–641.

Kolek, Erik (2024). On the technological foundations of interstellar space travel. In: *Chronicles of Business Informatics Physics (CBIP)*. Volume 3, edition no. 1.0. ISBN: 9783759785237.

Table of contents

In this research article, an advanced engineering method called Refit is developed for designing technological innovations. Refit consists of the phases definition, research, learning, working and product. These phases can be applied step by step and you can go one step forward or backward at any time. At the end of the development process, the definition phase is started again immediately.

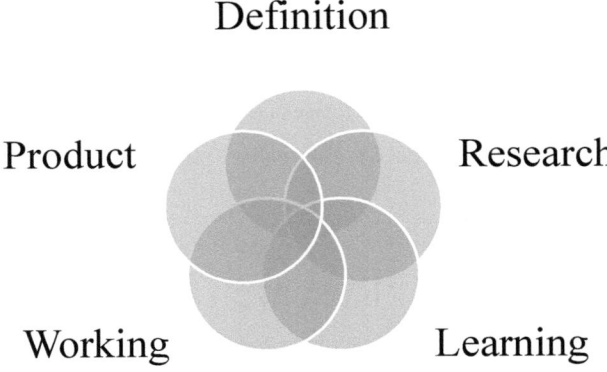

Figure 3. Table of contents illustrated by the importance of the refit technology development method.

Kolek, Erik (2024). On the technological foundations of interstellar space travel. In: *Chronicles of Business Informatics Physics (CBIP)*. Volume 3, edition no. 1.0. ISBN: 9783759785237.

Scientific citation:

Kolek, Erik (2024). The Erik-Kolek-Motor (E-KOMO) – An advanced modular, switchable electric generator motor. In: *On the technological foundations of interstellar space travel.* Chronicles of Business Informatics Physics (CBIP). Volume 3, edition no. 1.0.

Erik Kolek (2024)

The Erik-Kolek-Motor (E-KOMO) – An advanced modular, switchable electric generator motor

Inventor: Dr. rer. pol. Erik Kolek

Description

[0001] The invention is an advanced modular, switchable electric generator motor, i.e. a combination of at least one electric motor and at least one electric generator that can generate electricity inductively, while the electric motor absorbs electricity and provides the kinetic energy that is converted in the electric generator. The user can choose whether either more electric motors or more electric generators are actively connected electrically. This decision by the user depends solely on the desired volt output as locomotion power. Full volt output on all electric motors ensures the greatest acceleration, but also the lowest energy recovery and vice versa. The invention in the form of a modular and switchable electric generator motor is hereinafter referred to as the Erik-Kolek-Motor, or E-KOMO for short.

[0002] The E-KOMO is manufactured in the electrical engineering sector. Direct current or alternating current can be used to operate the E-KOMO. Only electromagnets make the E-KOMO cheaper to produce. Permanent magnets require rare earths such as iron, cobalt and nickel, so this type of magnet should generally be avoided. The E-KOMO is used in a variety of applications (in almost all appliances with or without power cables; be it cars, power stations, space travel, toy cars, toys,

Kolek, Erik (2024). On the technological foundations of interstellar space travel. In: *Chronicles of Business Informatics Physics (CBIP)*. Volume 3, edition no. 1.0. ISBN: 9783759785237.

lawnmowers, food processors, dishwashers, washing machines, electric shutters, escalators, trains, forklifts, refrigerators, production machines, etc.). 50 percent of the world's energy flows into motors and is consumed without being recovered. The E-KOMO solves the latter problem by always converting part of the kinetic energy into electricity. Of course, there will still be outdated conventional electric motors and traditional electric generators on the market. This means that there are not yet any comparable competitor products, as these only focus on energy efficiency and not on energy recovery in accordance with Albert Einstein's law of conservation of momentum for energy $E = mc^2$ [1].

[0003] The development of the E-KOMO follows the refit approach devised by Erik Kolek, which has its origins in Gene Roddenberry's well-known Star Trek reality. Any technology is selected, in this case the electric motor, and this technology is then improved step by step. The latter is done until a higher technology level or refit level has been reached at the end of development. Of course, the refit development method can be applied almost endlessly. What is important at the next refit stage is that the positive development aspects are always taken along and the negative development aspects are sorted out. The E-KOMO is a large refit. Smaller refits are, for example, individual components or technology learning aspects such as the eKolek-Newton pendulum (**Fig. 15**).

[0004] The E-KOMO is sustainable in its use of energy and corresponds to a green technology approach that focuses on and takes a closer look at the motor, for example of a car or other motor vehicle, instead of a charging infrastructure or battery. Economic, ecological and social aspects of electric mobility are thus taken into account. The E-KOMO saves money and time, is more energy-efficient than a conventional electric motor, causes correspondingly lower CO_2 emissions and is quieter and more powerful than combustion engines, as these do not recover any energy and do not have the potential to find widespread social acceptance in society. Users would like to have at least equivalent or better alternatives to the combustion

Kolek, Erik (2024). On the technological foundations of interstellar space travel. In: *Chronicles of Business Informatics Physics (CBIP)*. Volume 3, edition no. 1.0. ISBN: 9783759785237.

engine. The E-KOMO satisfies this social need by offering the same performance and even the possibility of recovering energy when idling or partially idling. To this end, the E-KOMO can be switched in different ways (**Fig. 1**). If at least two electric generator motors are mounted on the shaft, it can be selected according to the circuit diagram that (a) two electric motors are running, (b) one electric generator and one electric motor are on or (c) two electric generators are idling and generating energy. The latter is very easy to understand if you imagine, for example, a driven rear axle and a passive front axle of an automobile (**Fig. 2**). The E-KOMO also offers the option of placing more than just two electric generators or motors on one shaft, meaning that more electric motors or more electric generators can be freely switched as required. The advantages of a switchable electric generator motor are easy to understand on the E-KOMO (**Fig. 1**). It is important that the components of the electric generator motor can be switched individually, as groups of armatures (**3, 13, 14**) on commutators (**1**) would make the E-KOMO more inefficient and cumbersome. This should of course be avoided.

[0005] The technologically innovative advantages of the E-KOMO are easy to recognize when you look at its possible applications in cars, for example (**Fig. 2**). A heavy gearbox is no longer necessary, as the E-KOMO can be installed directly on the front and rear axle. Instead of the usual gearshift, the desired speed setting is conveniently determined by the user using the pedal, just like the acceleration of a combustion engine. E-motors are switched on or off accordingly to ensure energy recovery. Today, more gas means more volts; above a certain voltage, electric motors automatically switch on or electric generators switch off. As soon as you stop accelerating, the battery is recharged while you are driving and not just at the charging station, for example at home. For charging stations, an external power connection to the distributor must also be taken into account. The latter is superfluous as soon as a photovoltaic system is installed on the roof to support the E-KOMO. A connection to the inverter must be added for this, so that the battery can be charged at any time regardless of the location of the vehicle. At the moment, the rear-wheel drive is on in

Kolek, Erik (2024). On the technological foundations of interstellar space travel. In: *Chronicles of Business Informatics Physics (CBIP)*. Volume 3, edition no. 1.0. ISBN: 9783759785237.

the circuit diagram, while energy is only produced at the front and fed to the power electronics (**Fig. 2**). This enables a much greater range, also because the car is ultra-light without the additional weight of the transmission and combustion engine. The car is energy-efficient and yet sporty in its acceleration. Thanks to the E-KOMO, the car of the future of the 21st century is possible.

[0006] In the following design variants, the motto is always: what you see is what you get. This means that experiences are explained so that the subsequent design variants are easier to understand. All of the design variants shown are functioning illustrative models and will later be in industrial production in a housing. Ultimately, the aim is to further develop the electric motor invented by Thomas Davenport thanks to the technology innovation method Refit. Accordingly, further improvements can be made directly during the subsequent industrial implementation. This is only about the invention and claims to protection arise from the further development of the electric motor into an electric generator motor or E-KOMO (also given by copyrights).

[0007] **Design variant 1.** Free-swinging permanent magnets (**4, 5**) increase the performance of the E-KOMO (**Fig. 3** and **Fig. 4**). Although a free-moving magnetic field (**4, 5**) increased the performance, the E-KOMO proved to be somewhat more unsteady and less stable overall. For this reason, this design variant was not researched more extensively (apart from the suggestion, as below, to make better use of this effect in space travel). Possibilities that could arise from this may be better understood in the vacuum of our universe. Until then, the circuit between at least one e-motor on the one hand and at least one e-generator on the other remains thanks to the inductive electrical principle. The simplest E-KOMO consists of an e-motor and an e-generator while current is conducted through the e-motor (**Fig. 3**). Accordingly, the E-KOMO becomes more stable in operation and more powerful when two components are doubled, as is the case here with the armatures (**8**), each of which is attached to a commutator (**1**) (**Fig. 4**). In general, the vibrations were damped by installing a connecting strut (**10**) above to guide the (permanent) magnets (**4**). These are double-pole armatures **3**. Double armatures (**8**) on a commutator (**1**) can be seen in **Fig. 4**.

[0008] The observed magnetic effect of the E-KOMO describes a frequency-based interaction between magnetic fields according to electrodynamics. In experiments with (electric) magnets, a mutual oscillation amplitude is visible, which further increases the performance of the E-KOMO in terms of efficiency due to the additional magnetic kinematic energy gained, as well as all other conventional electric motors and generators; also in power plants. This is a principle on the basis of which a more advanced electrodynamics is created, the electromagnetic dynamics. In the vacuum of our universe, the observed magnetic effect is possibly the basis for a magnetic impulse drive which, with a correspondingly long acceleration, could make it possible to approach the speed of light in an energy-efficient manner (**Fig. 20**); however, in research practice, the constructed magnetic impulse drive is certainly only an energy-efficient maneuvering drive.

[0009] **Design variant 2.** Thanks to the learned magnetic effects, free-swinging magnets (**4, 5**) were dispensed with (**Fig. 5** and **Fig. 6**). This resulted in smooth-running and stable electric motors and generators. The E-KOMO can be freely switched back and forth (**Fig. 1** and **Fig. 5**). The special feature is the modular design, which means that the E-KOMO can be installed one after the other as often as required (**Fig. 6**). In any case, there is always a direct connection to the shaft or drive axle (**6**).

[0010] **Design variant 3.** The permanent magnets in the E-KOMO have been replaced by electromagnets (**4**) in order to take ecological aspects of sustainability more into account. Furthermore, there is switchability (**11**) in accordance with the intended circuit diagram (**Fig. 1**, **Fig. 7** and **Fig. 8**). The use of the double armature (**8**) made the E-KOMO in **Fig. 8** more powerful and lively, but also somewhat more unsteady and slower compared to the variant shown in **Fig. 9**. Here the E-KOMO ran like clockwork and its advantages in the circuit (**11**) came into their own. It is important to ensure that the e-magnets are permanently energized, otherwise no energy can be

Kolek, Erik (2024). On the technological foundations of interstellar space travel. In: *Chronicles of Business Informatics Physics (CBIP)*. Volume 3, edition no. 1.0. ISBN: 9783759785237.

recovered. The best way to do this is to switch each commutator (**1**) individually (**11**). These are also two-pole armatures (**3**).

[0011] **Design variant 4.** To further accelerate the running performance of the E-KOMO, the number of poles was increased to three (**13**) and four poles (**14**) (**Fig. 9** to **Fig. 12**). It turned out that the higher the number of poles, the more efficient and faster the E-KOMO works. The size also plays a role here, because the smaller the armature (**3, 13, 14**), the faster it seems to be able to rotate, but it is weaker in terms of overcoming friction, for example in a stationary car. This means that a smaller E-KOMO does not necessarily have to have less power. The arrangement of the permanent magnets or electromagnets on the side (**9**) (**Fig. 9** and **Fig. 10**) is interesting, as this results in double the acceleration. Accordingly, armatures (**3, 13, 14**) must be completely enclosed by electromagnets (**4**) in order to achieve the best possible performance. For this reason, an enclosing arrangement of the electromagnets (**4**) was again selected in **Fig. 11** and **Fig. 12**. The special feature is the shifting capability, because this time double anchors (**8**) were dispensed with and each anchor (**3, 13, 14**) was assigned to a commutator (**1**), which made the E-KOMO much more efficient and faster when shifting (**11**). Double anchors (**8**) are a good idea, but somewhat cumbersome to implement, but more powerful. The latter depends on what the E-KOMO is to be used for. In an automobile, for example, the armature (**3, 13, 14**) with commutator (**1**) is certainly more interesting for users because it allows more gears to be shifted. The E-KOMO seen in **Fig. 12** therefore runs very sportily, almost like a boxer engine.

[0012] **Fig. 13** shows the e-Kolek-Newton pendulum to explain inductive energy recovery consisting of kinetic energy and electrical power. It is based on the Newton pendulum and can be used individually in any machine in which motion has to be converted back into electrical energy. For example, in a car that starts and brakes, which means that every change in speed results in a certain amount of energy recovery, simply due to the e-Kolek-Newton pendulum used, which passes over

Kolek, Erik (2024). On the technological foundations of interstellar space travel. In: *Chronicles of Business Informatics Physics (CBIP)*. Volume 3, edition no. 1.0. ISBN: 9783759785237.

permanent magnets, or rather e-magnets, when oscillating. Other particularly interesting application examples here are trains and power stations, especially hydroelectric power stations, because every oscillation also means energy recovery. Ultimately, it was learned that a magnetic field must always be present to generate energy, preferably a combination consisting of a permanent magnet and an electromagnet.

[0013] **Design variant 5.** The number of poles (**13, 14**) was increased and permanent and electromagnets (**4**) were used to accelerate the running performance of the E-KOMO in a more energy-efficient manner (**Fig. 14** to **Fig. 16**). The combination of permanent and electromagnetization (**4**) has the advantage that you can choose to accelerate the E-KOMO thanks to the electric magnet and/or rely on the permanent magnet. Once again, a commutator (**1**) was assigned to each armature (**13, 14**) in order to guarantee smooth, stable and sporty running of the E-KOMO. **Fig. 16** shows a three-speed circuit (**11**), which, in contrast to the two-speed circuit (**11**) and four-speed circuit (**11**), represents the suitable circuit in the middle. Three gears appear to be suitable for two-wheeled machines such as e-bikes, e-scooters and e-motorcycles.

[0014] **Design variant 6.** The E-KOMO can be used in other variants (**Fig. 17** to **Fig. 19**). It can be built as a double or multi-dimensional roller motor or reed contact motor (**Fig. 17**). However, an even higher number of e-magnets is better, e.g. left and right as well as top and bottom, or mounted on all sides like a star. Connecting the E-KOMO to photovoltaic or solar cells (**19**) is possible at any time and makes sense (**Fig. 18**). The latter can even significantly improve existing solar power plants as soon as the E-KOMO is used there. Once again, the versatility of the E-KOMO is evident. It can also be implemented as a Stirling engine in which the hot air is generated via a simple copper winding (**20**) (**Fig. 19**).

[0015] **Design variant 7.** The E-KOMO is the basis for a magnetic impulse drive in the vacuum of our universe (**Fig. 20**). The special feature is its multidimensional rotatability, because on the one hand the anchors (**3**) rotate around the commutators

Kolek, Erik (2024). On the technological foundations of interstellar space travel. In: *Chronicles of Business Informatics Physics (CBIP)*. Volume 3, edition no. 1.0. ISBN: 9783759785237.

(1) and on the other hand the entire construction rotates around the anchors (3) and commutators (1). This magnetic impulse drive (or maneuvering drive) requires further research in the vacuum of our universe.

Protection claims of the E-KOMO which generally improve the familiar classic e-motor or e-generator

1. Arbitrarily selectable circuits between the electric motors or electric generators enable energy-efficient, sustainable use (**Fig. 1**). When idling, only electric generators are on, and all electric motors are on under full load. Accordingly, a certain number of motors and generators can be electrified in between according to the user's individual volt output requirements. The individual switchability of the armatures and the number of switches is important: two, three and four etc. Switching.

2. Free-swinging permanent magnets or correspondingly attached electromagnets can positively influence or increase the performance of the E-KOMO (e-generator motor) (oscillation frequency), but this **free oscillation of the magnets** must not be too large. In space travel, on the other hand, an appropriately constructed magnetic impulse drive could even offer the possibility of skillfully and precisely arranging smaller course corrections by shifting the point mass (maneuvering drive) without expending a great deal of energy.

3. Usability, especially in automobiles and similar machines (**Fig. 2**), thanks to the **modular design** of the E-KOMO.

4. At least two to four or more electric motors or generators are rotatably mounted on a **common shaft** or drive axle.

5. Split commutators to which two anchors are attached, at least one on the left and one on the right (so-called **double anchor 8**). Multiple anchors can be installed accordingly.

6. The **e-Kolek-Newton pendulum** as a new way of recovering energy from kinetic energy and converting it back into electrical power.

7. The continuous most efficient use of two types of magnets simultaneously: electromagnets and permanent magnets are always to be installed simultaneously in order to obtain the best electrical induction performance for energy recovery (**characterized by** claim 6). The parallel connection of the electromagnets connected to the individually switchable commutators, to which a two- to x-pole armature is connected.

8. The **design variants** of the E-KOMO shown in **Fig. 3** to **Fig. 20** and the electrical circuit diagrams shown therein.

9. All **related embodiments** of the E-KOMO of industrial type (**characterized by** claim 8).

10. Multidimensionality of the modular design thanks to several electromagnets in the roller motor or reed contact motor (**Fig. 17**).

11. The **combination** of E-KOMO with photovoltaic or solar cells to increase the efficiency of corresponding power plants.

12. Generation of hot air by copper windings in the Stirling-E-KOMO.

13. The magnetic impulse drive as a maneuvering drive thanks to the E-KOMO (**Fig. 20**).

List of reference symbols

1	Commutator
2	Copper contact / iron contact
3	Armature (rotor) two-pole
4	Permanent magnet top / electromagnet top

Kolek, Erik (2024). On the technological foundations of interstellar space travel. In: *Chronicles of Business Informatics Physics (CBIP)*. Volume 3, edition no. 1.0. ISBN: 9783759785237.

5	Free-swinging bracket
6	Motor shaft / motor axle
7	Permanent magnet bottom / electromagnet bottom
8	Double armature (double rotor) two-pole on one commutator
9	Lateral permanent magnet / Lateral electromagnet
10	Fixed bracket
11	Circuit of the commutator
12	Parallel connection
13	Armature (rotor) three-pole
14	Armature (rotor) four-pole
15	Free-swinging electromagnet(s)
16	Rotating permanent magnet(s)
17	Reed contact
18	Fixed electromagnet
19	Photovoltaic / solar cell
20	Copper winding

This is followed by 19 pages of illustrations.

Kolek, Erik (2024). On the technological foundations of interstellar space travel. In: *Chronicles of Business Informatics Physics (CBIP)*. Volume 3, edition no. 1.0. ISBN: 9783759785237.

References

[1]   A. Einstein (1905). Does the Inertia of a Body Depend upon its Energy Content? *Annalen der Physik 18(13)*, pp. 639–641.

Kolek, Erik (2024). On the technological foundations of interstellar space travel. In: *Chronicles of Business Informatics Physics (CBIP)*. Volume 3, edition no. 1.0. ISBN: 9783759785237.

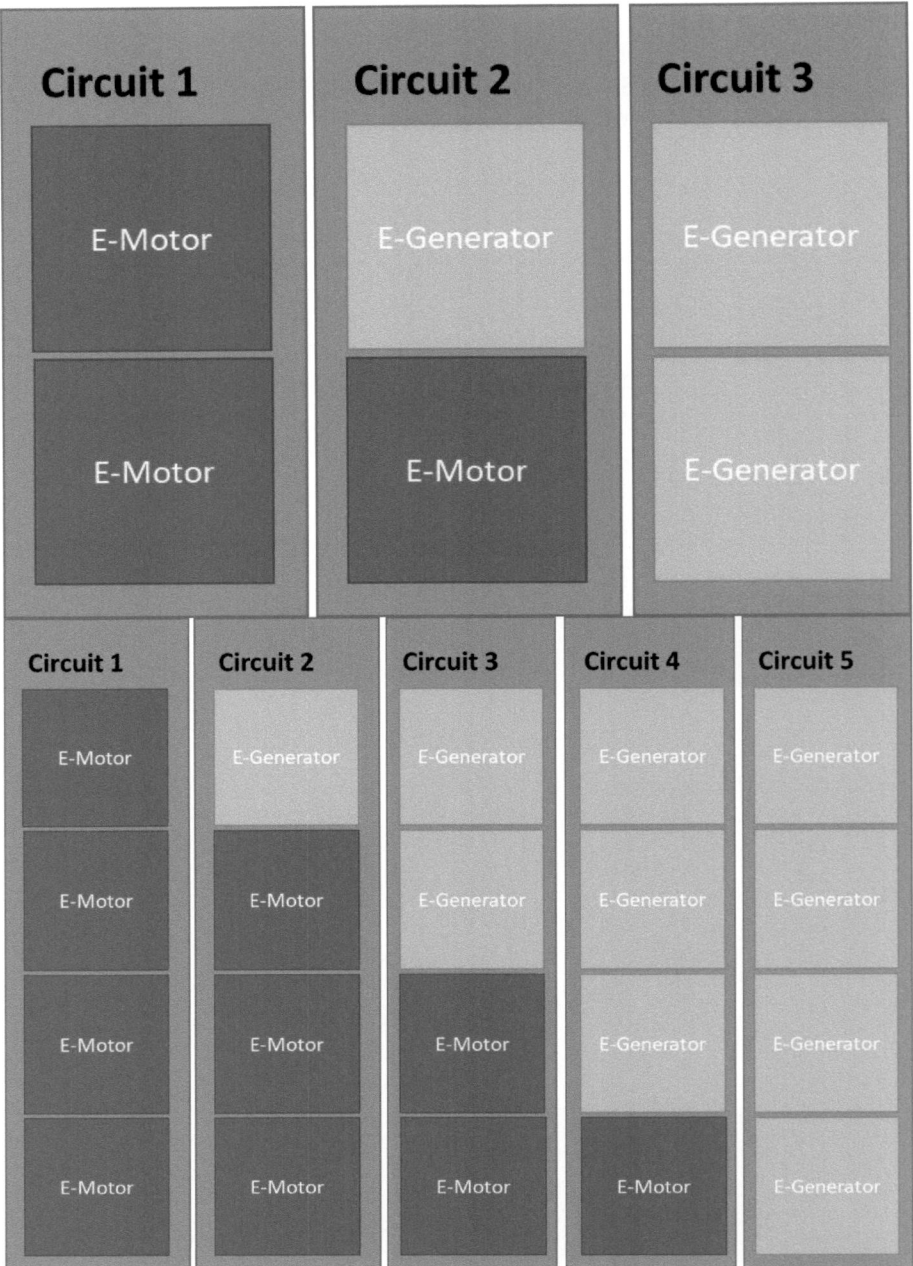

Fig. 1. Two circuit diagrams of the E-KOMO with two or four electric generator motors.

Kolek, Erik (2024). On the technological foundations of interstellar space travel. In: *Chronicles of Business Informatics Physics (CBIP)*. Volume 3, edition no. 1.0. ISBN: 9783759785237.

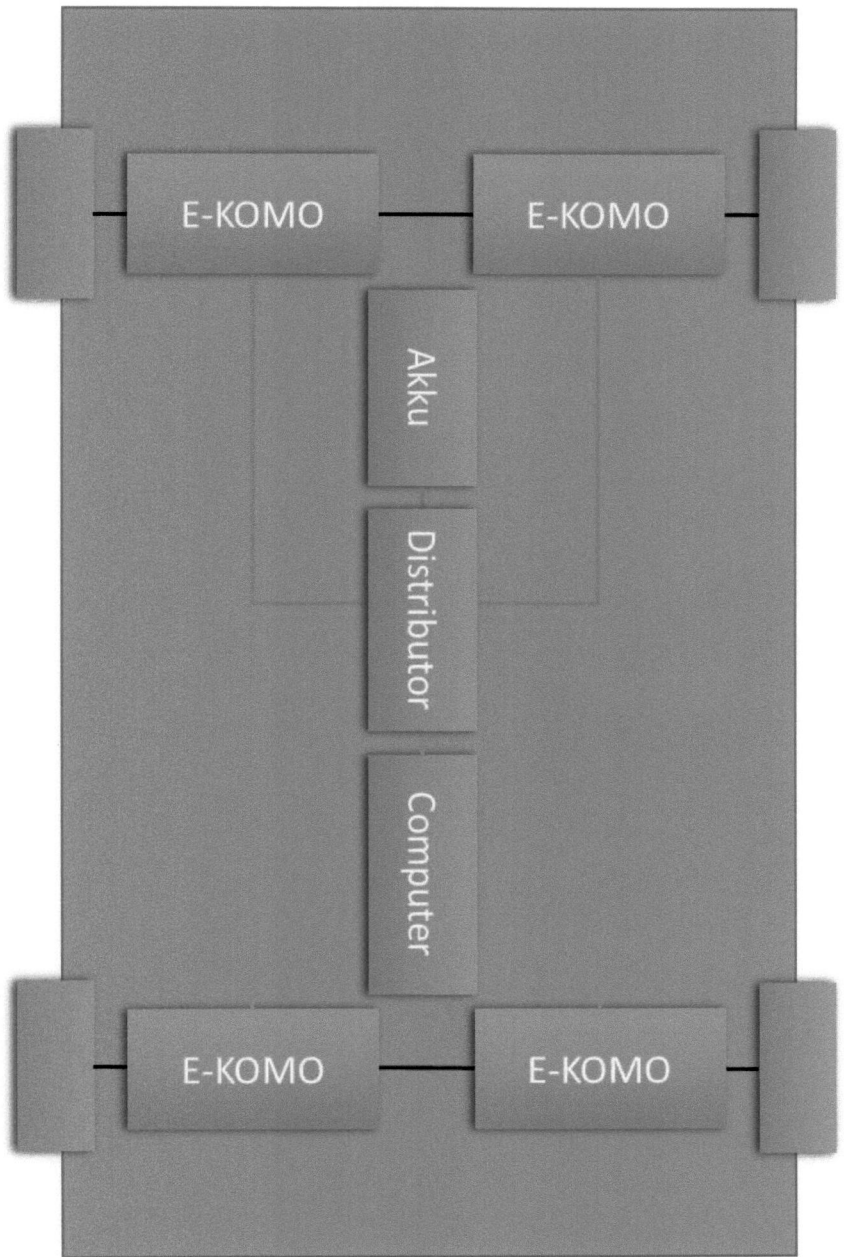

Fig. 2. Circuit diagram of the electrical circuits of the E-KOMO using the example of the automobile.

Kolek, Erik (2024). On the technological foundations of interstellar space travel. In: *Chronicles of Business Informatics Physics (CBIP)*. Volume 3, edition no. 1.0. ISBN: 9783759785237.

Fig. 3. The E-KOMO in modular design with free-swinging permanent magnets and two armatures on one commutator each.

Kolek, Erik (2024). On the technological foundations of interstellar space travel. In: *Chronicles of Business Informatics Physics (CBIP)*. Volume 3, edition no. 1.0. ISBN: 9783759785237.

Fig. 4. The E-KOMO in modular design with free-swinging permanent magnets and two double armatures, each with a split commutator.

Kolek, Erik (2024). On the technological foundations of interstellar space travel. In: *Chronicles of Business Informatics Physics (CBIP)*. Volume 3, edition no. 1.0. ISBN: 9783759785237.

Fig. 5. The E-KOMO in modular design with fixed permanent magnets and an armature on each commutator.

Kolek, Erik (2024). On the technological foundations of interstellar space travel. In: *Chronicles of Business Informatics Physics (CBIP)*. Volume 3, edition no. 1.0. ISBN: 9783759785237.

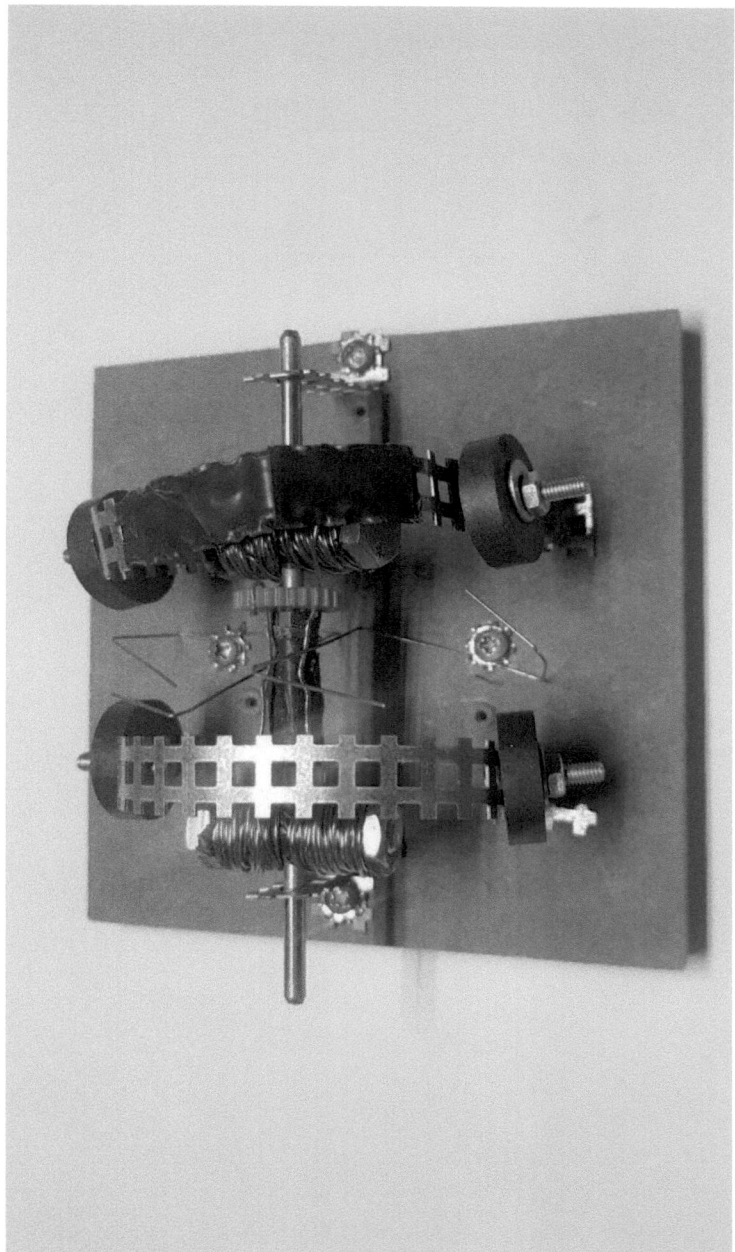

Fig. 6. The E-KOMO in modular design with fixed permanent magnets and double armature on a split commutator.

Kolek, Erik (2024). On the technological foundations of interstellar space travel. In: *Chronicles of Business Informatics Physics (CBIP)*. Volume 3, edition no. 1.0. ISBN: 9783759785237.

Fig. 7. The E-KOMO in modular design with permanently installed electromagnets and armatures on one commutator each for freely selectable switchability.

Kolek, Erik (2024). On the technological foundations of interstellar space travel. In: *Chronicles of Business Informatics Physics (CBIP)*. Volume 3, edition no. 1.0. ISBN: 9783759785237.

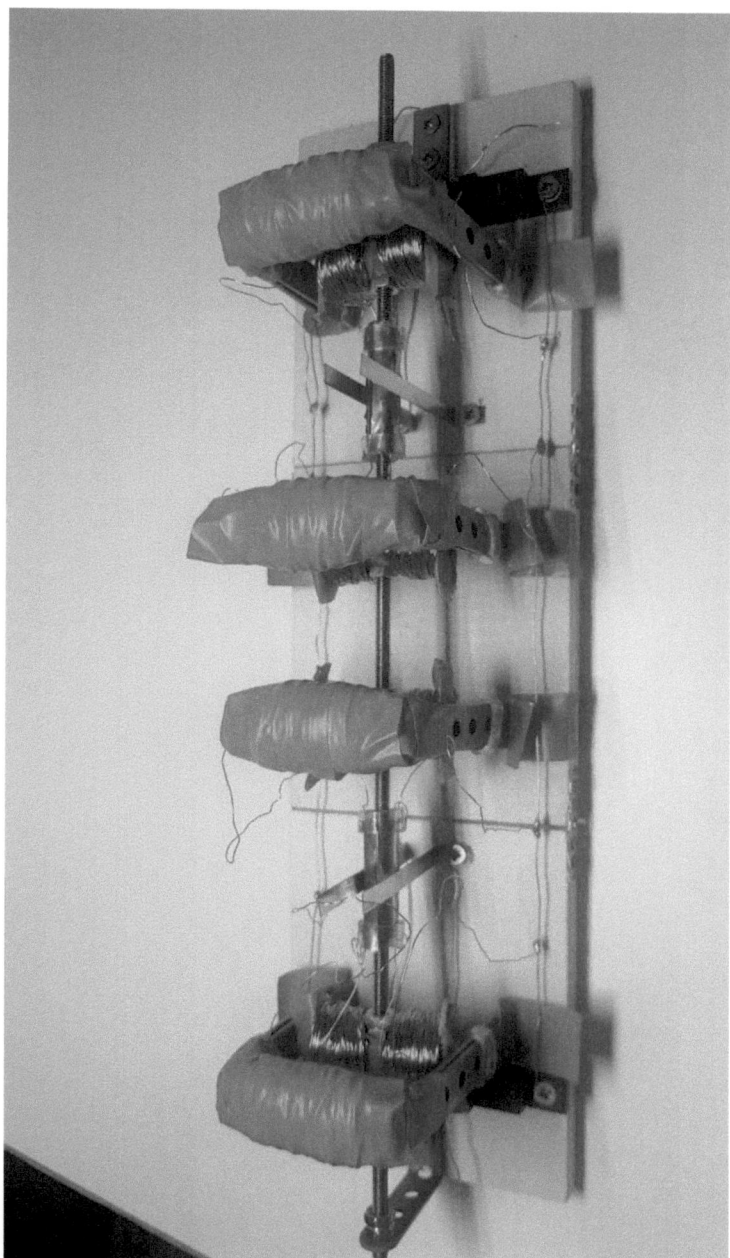

Fig. 8. The E-KOMO in modular design with permanently installed electromagnets and double armatures on one commutator each for freely selectable switchability.

Kolek, Erik (2024). On the technological foundations of interstellar space travel. In: *Chronicles of Business Informatics Physics (CBIP)*. Volume 3, edition no. 1.0. ISBN: 9783759785237.

Fig. 9. The E-KOMO in modular design with permanently installed permanent magnets and three-pole armatures on one commutator each for freely selectable switchability.

Kolek, Erik (2024). On the technological foundations of interstellar space travel. In: *Chronicles of Business Informatics Physics (CBIP)*. Volume 3, edition no. 1.0. ISBN: 9783759785237.

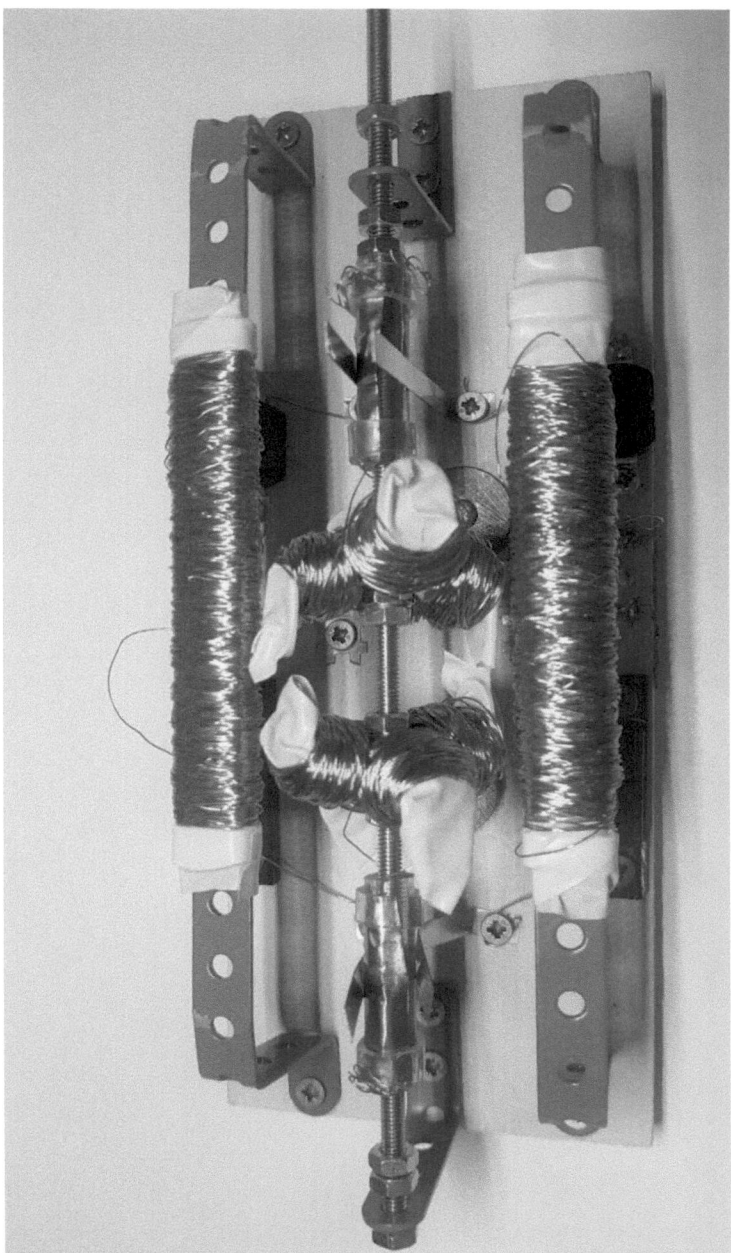

Fig. 10. The E-KOMO in modular design with fixed electromagnets on the side and four-pole armatures on one commutator each for freely selectable switchability.

Kolek, Erik (2024). On the technological foundations of interstellar space travel. In: *Chronicles of Business Informatics Physics (CBIP)*. Volume 3, edition no. 1.0. ISBN: 9783759785237.

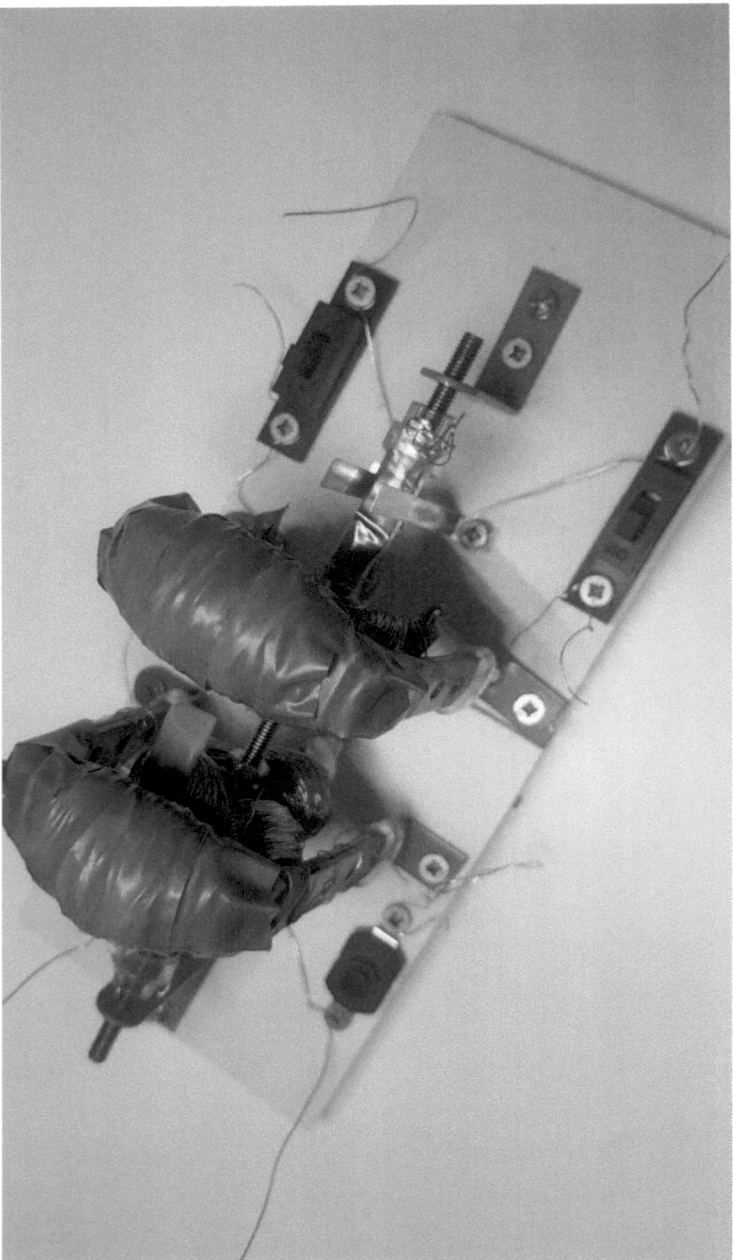

Fig. 11. The E-KOMO in modular design with fixed electromagnets above and two four-pole armatures on one commutator each for freely selectable switchability.

Kolek, Erik (2024). On the technological foundations of interstellar space travel. In: *Chronicles of Business Informatics Physics (CBIP)*. Volume 3, edition no. 1.0. ISBN: 9783759785237.

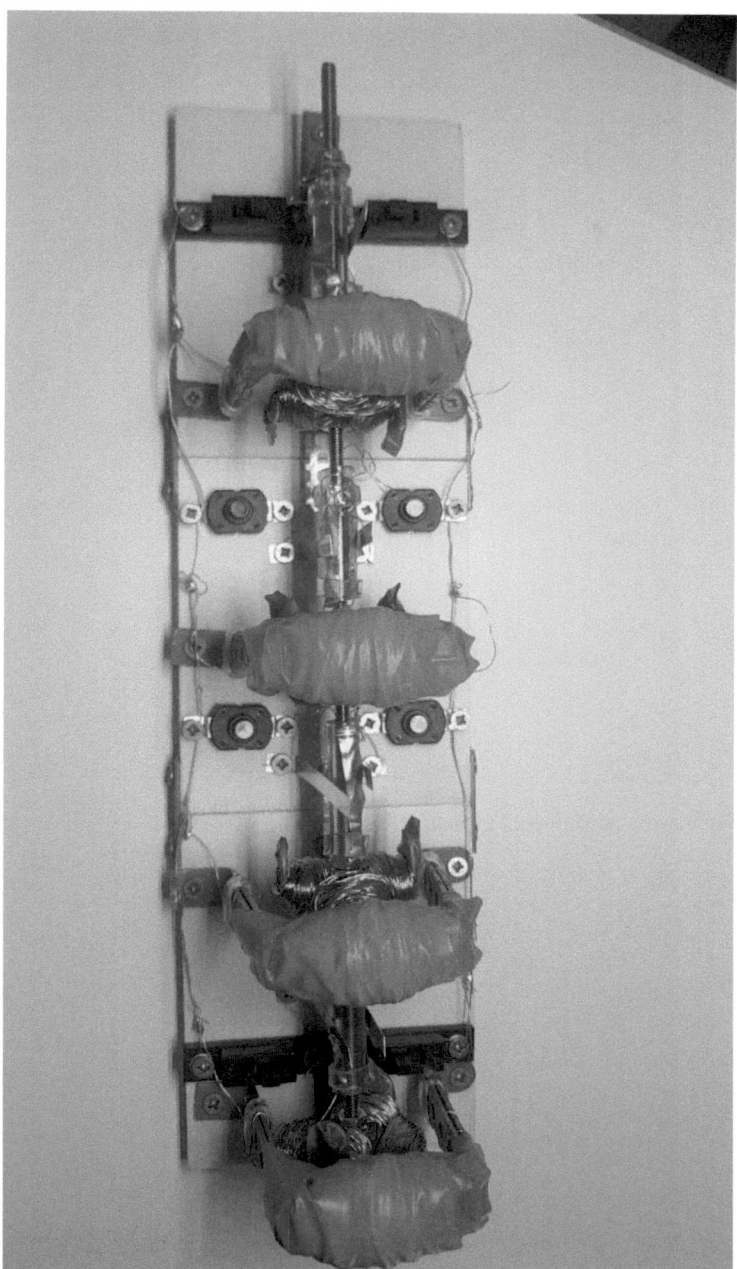

Fig. 12. The E-KOMO in a modular design with electromagnets permanently installed above and four four-pole armatures on one commutator each for freely selectable switchability.

Kolek, Erik (2024). On the technological foundations of interstellar space travel. In: *Chronicles of Business Informatics Physics (CBIP)*. Volume 3, edition no. 1.0. ISBN: 9783759785237.

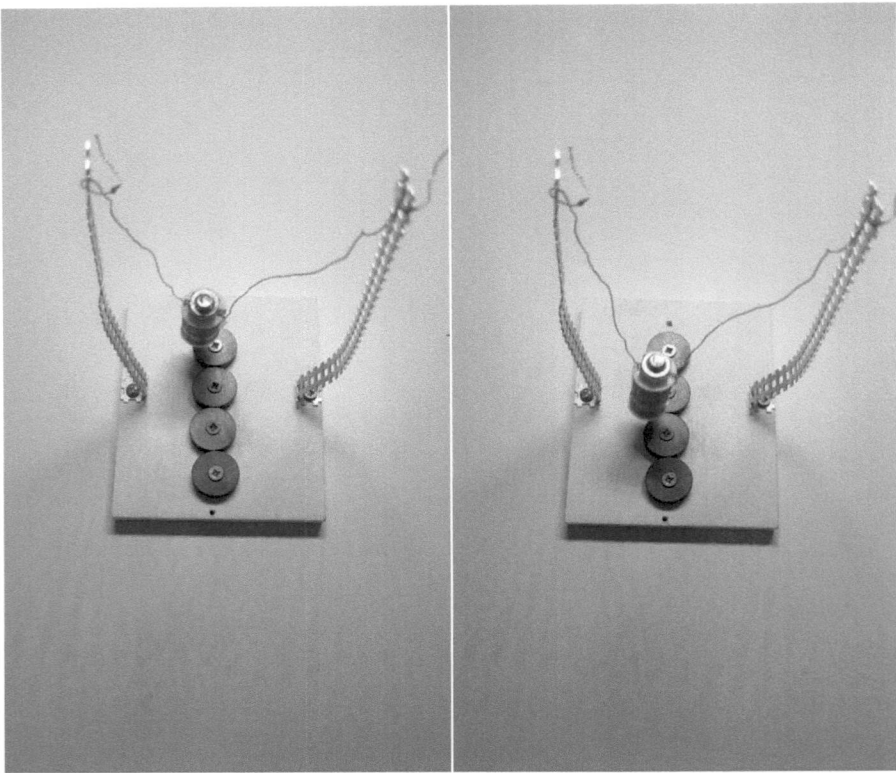

Fig. 13. The e-Kolek-Newton pendulum to explain inductive energy recovery.

Kolek, Erik (2024). On the technological foundations of interstellar space travel. In: *Chronicles of Business Informatics Physics (CBIP)*. Volume 3, edition no. 1.0. ISBN: 9783759785237.

Fig. 14. The E-KOMO in modular design with permanently installed permanent and electromagnets and three-pole armatures on one commutator each for freely selectable switchability.

Kolek, Erik (2024). On the technological foundations of interstellar space travel. In: *Chronicles of Business Informatics Physics (CBIP)*. Volume 3, edition no. 1.0. ISBN: 9783759785237.

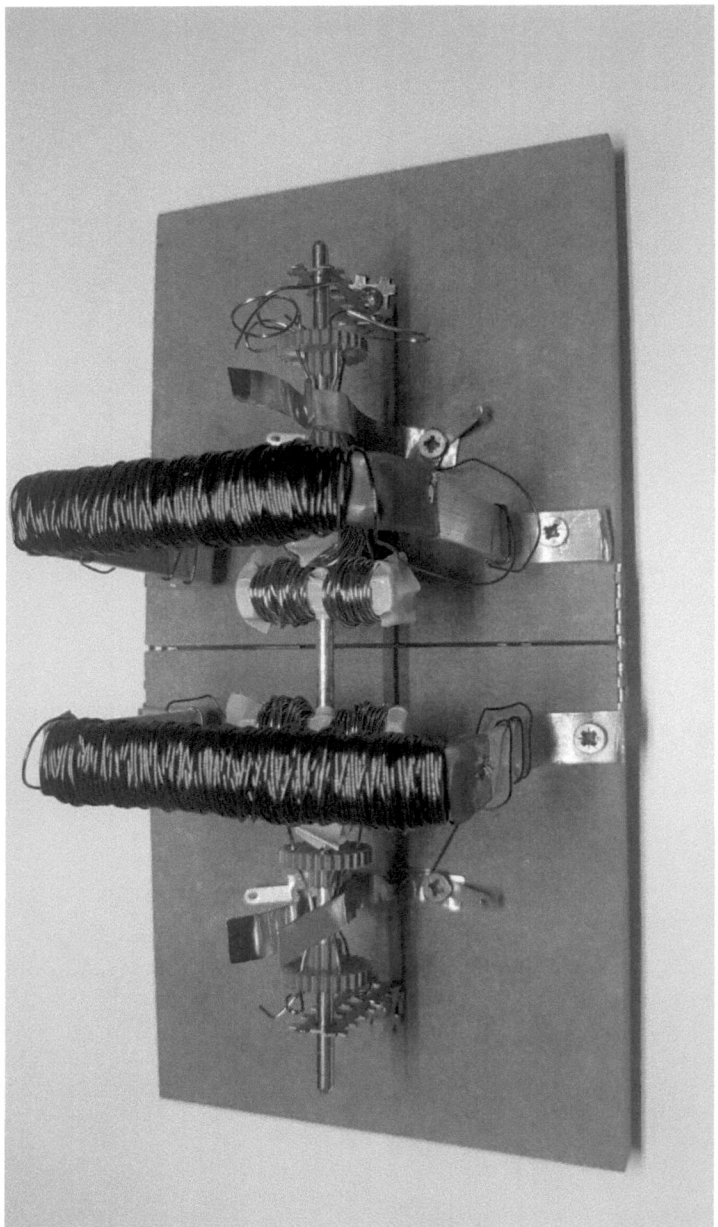

Fig. 15. The E-KOMO in modular design with permanently installed permanent magnets and electromagnets and two four-pole armatures, each on a commutator for freely selectable switchability.

Kolek, Erik (2024). On the technological foundations of interstellar space travel. In: *Chronicles of Business Informatics Physics (CBIP)*. Volume 3, edition no. 1.0. ISBN: 9783759785237.

Fig. 16. The E-KOMO in a modular design with permanently installed permanent and electromagnets and three four-pole armatures, each on a commutator for freely selectable switchability.

Kolek, Erik (2024). On the technological foundations of interstellar space travel. In: *Chronicles of Business Informatics Physics (CBIP)*. Volume 3, edition no. 1.0. ISBN: 9783759785237.

Fig. 17. The E-KOMO in modular design with permanently installed permanent and electromagnets for freely selectable switchability.

Kolek, Erik (2024). On the technological foundations of interstellar space travel. In: *Chronicles of Business Informatics Physics (CBIP)*. Volume 3, edition no. 1.0. ISBN: 9783759785237.

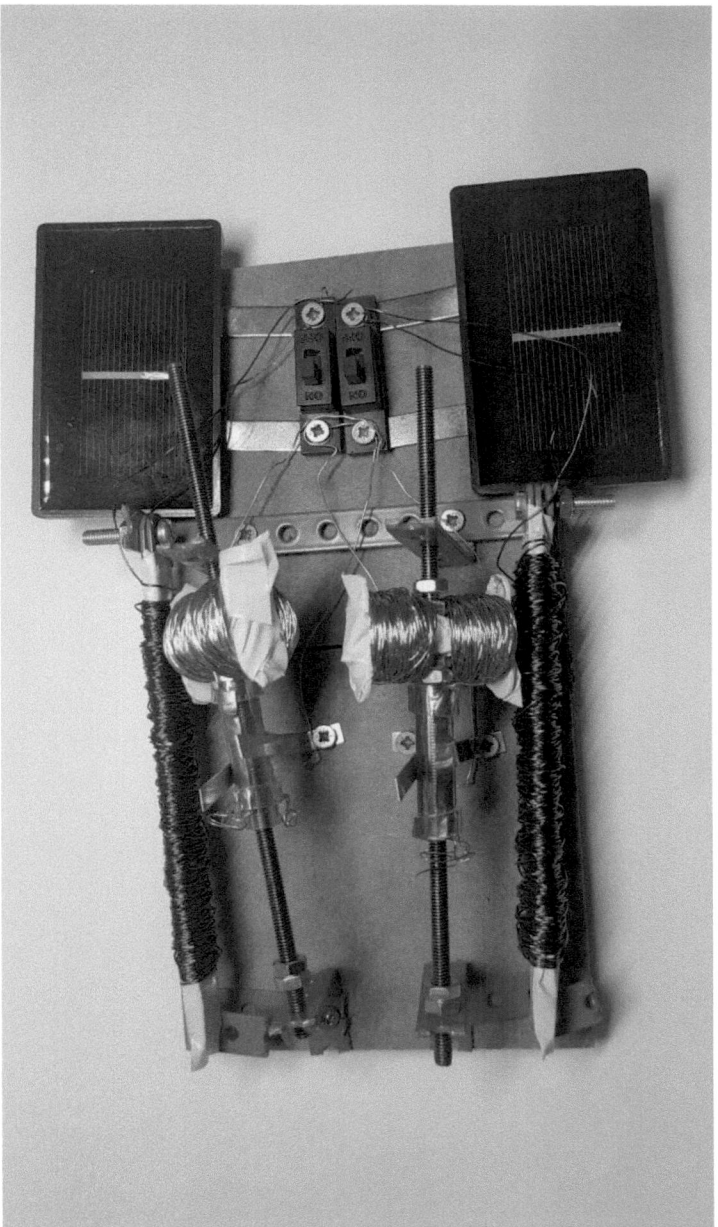

Fig. 18. The E-KOMO in modular design with permanently installed electromagnets and two-pole armatures on one commutator each for arbitrarily selectable switchability and connected to a photovoltaic system.

Kolek, Erik (2024). On the technological foundations of interstellar space travel. In: *Chronicles of Business Informatics Physics (CBIP)*. Volume 3, edition no. 1.0. ISBN: 9783759785237.

Fig. 19. The E-KOMO in modular design with permanently installed electromagnets for generating hot air in the Stirling engine.

Kolek, Erik (2024). On the technological foundations of interstellar space travel. In: *Chronicles of Business Informatics Physics (CBIP)*. Volume 3, edition no. 1.0. ISBN: 9783759785237.

Fig. 20. The E-KOMO in modular design with permanently installed electromagnets and two-pole double armatures on one commutator each for arbitrarily selectable switchability and with self-turning option of the ring socket.